Life of Fred®

Pre-Algebra 1 with Biology

Life of Fred®

Pre-Algebra 1 with Biology

Stanley F. Schmidt, Ph.D.

Polka Dot Publishing

© 2015 Stanley F. Schmidt
All rights reserved.

ISBN: 978-0-9791072-2-1

Library of Congress Catalog Number: 2009937219
Printed and bound in the United States of America

Polka Dot Publishing   Reno, Nevada

To order copies of books in the Life of Fred series,

visit our Web site PolkaDotPublishing.com

Questions or comments? Email the author at lifeoffred@yahoo.com

Seventh printing

Life of Fred: Pre-Algebra 1 with Biology was illustrated by the author with additional clip art furnished under license from Nova Development Corporation, which holds the copyright to that art.

for Goodness' sake

or as J.S. Bach—who was never noted for his plain English—often expressed it:

Ad Majorem Dei Gloriam
(to the greater glory of God)

If you happen to spot an error that the author, the publisher, and the printer missed, please let us know with an email to: lifeoffred@yahoo.com

As a reward, we'll email back to you a list of all the corrections that readers have reported.

Pre-Algebra

Is there a time between childhood and adulthood? Yes and no.

Yes. There are those million adolescent years that begin when you first start to get hair under your arms and ends somewhere between ages 18 and 28. My daughter Margaret once told me that she wanted to "grow up and be a dolt like you." She was three when she said that. She became an English major in college.

No. There is no time between childhood and adulthood. There are years in which childhood and adulthood are blended together. These are years in which we hold onto parts of childhood with one hand as we stretch out to grasp parts of adulthood with the other hand. In those years, we sometimes make the noises that kids make. And sometimes we make the noises that adults make.

Is there such a thing as pre-algebra? Yes and no.

Yes. The book that you are holding is a pre-algebra book. Doesn't that prove that pre-algebra exists?* If you tell your grandma that you are studying pre-algebra, she will pat you on the head and say, "I'm proud of you." In contrast, if you tell your grandma that you have a pet unicorn, she will tell you, "You are a big silly." She will pat you on the head and feel for soft spots.**

---

* Of course, if there were a *Life of Fred: Horses With Wings* book, would that mean that horses with wings exist?

** Since this is a biology book, this is a good time to start. Your grandma is feeling your head to find soft spots on your head called **fontanels**. When you become an adult, they may call you a bonehead, but babies at birth might be called boneheads. Their skulls have about seven bones. (That makes being born a lot easier.)

For many first-time parents, they get nervous when they notice the fontanels on their newborn. There's a big soft spot in front and a much smaller one in back. (When Fred was born, it was his nose that scared his parents.) The fontanels disappear in the first two years of life as the bones fuse together.

No. There is no such thing as pre-algebra. I checked two dictionaries and "pre-algebra" wasn't in either one of them.*

A better argument for the non-existence of pre-algebra is that there is no mathematics that is uniquely pre-algebra. It is just a blend of arithmetic and the first parts of algebra.

"Solve 2 + 2" is arithmetic.
"Solve $\frac{2}{3} + \frac{4}{5}$" is arithmetic.
"8 is what percent of 32?" is arithmetic. $\Big\}$ are all called pre-algebra.
"Solve 3x + 6 = 5x" is algebra.
"Combine $x^5 x^4$" is algebra.

So why did they invent the name "pre-algebra"? For the same reason they invented the name "teenager." It gives a name to the place you stand when you are between two worlds.

childhood ↔ adulthood
arithmetic ↔ algebra

---

* Is this a good argument? I'm not in the dictionary. You are not in the dictionary. But we exist.

To make things worse, *unicorn* is in both of my dictionaries. And in Deuteronomy 33:17 (King James Version).

8

Biology

You go to school. You sit in a room for fifty minutes and study history. Then they ring a bell, and you go into another room and study a foreign language. Then they ring a bell, and you go into another room and study biology.

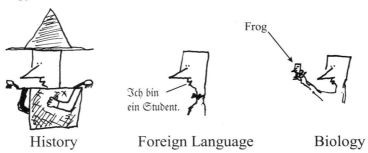

History Foreign Language Biology

Doesn't that seem a little unnatural to you? The world is not divided up into little water-tight compartments of aerospace science & engineering, agricultural & environmental education, animal biology, animal science & management, anthropology, art history, art studio, atmospheric science, avian sciences, biochemical engineering, biochemistry & molecular biology, biological sciences, biological systems engineering, biomedical engineering, biotechnology, cell biology, chemical engineering, chemistry, Chinese, civil engineering, classical civilization, clinical nutrition, communication, comparative literature, computer engineering, computer science, dramatic art, ecological management & restoration, economics, electrical engineering, English, entomology, exercise biology, fiber & polymer science, film studies, food science, French, genetics, geology, German, history, human development, hydrology, international agricultural development, international relations, Italian, Japanese, landscape architecture, linguistics, *mathematics*, mechanical engineering, medieval & early modern studies, microbiology, music, neurobiology, physiology, nutrition science, optical science & engineering, philosophy, physics, plant biology, political science, psychology, religious studies, Russian, sociology, Spanish, statistics, textiles & clothing.

One teacher will say, "I teach chemical engineering."
Another will say, "I teach German."
Let us give five stars to the teacher that says, "I teach students." What a novel idea!—treating students as if they were humans, rather than receptacles into which they can pour a single distilled subject.

So this is a pre-algebra book . . . and a biology book. And we will quote six lines of Italian poetry. We will discuss the difference between a metaphor and a simile. Talk about first-aid for fainting. Balance some chemical equations. Play with a French phrase in the movie *Camelot*.

And in the meantime, you will learn some algebra and a lot of interesting biology.

INTERESTING BIOLOGY

Is "interesting biology" an oxymoron?* Your typical modern biology book weighs a ton and costs a fortune. I opened one of my biology books at random. The heading was "Producers of Zygosporangia." The first sentence under that heading was "Consider the zygomycetes." A second biology book has a whole chapter entitled "Excretion." No thanks.

One reason I chose to become a mathematician rather than a biologist is the mind-numbing number of terms that you have to memorize in beginning biology. Did I want to learn that insects get rid of water through Malpighian tubules? Or, flipping to another page in another biology book, we learn, "11-*cis* retinaldehyde combines with a protein, opsin, to form the visual pigment rhodopsin."

Names, names, names. If I know the names of every frog in the pond, do I really *know* anything?

But getting beyond all the memorizing of details, there is a ton of interesting parts of biology—several ounces of which are in this book. You learn why, if George Washington had traveled to every place in the whole world, he would never have found a Concord grape. (You'll also learn how a two-year-old, Sally, became the owner of an 80-store mall, and about silver wedding axes—but neither of these have anything to do with biology.)

Many students who have had a year of high school biology can't tell the difference between a gene and a chromosome.

Or they look at a six-ton tree. They know that the water in the tree came from the roots. You ask them, "Where did the rest of the mass of the tree come from?" Many don't know. (It did *not* come from the dirt.) You will find out in this book.

---

* OX-ee-MORE-on. An oxymoron is a phrase that seems to be contradictory, like "painless IRS audit" or "full-length bikini."

About This Book

Generally speaking, each chapter is a lesson. Some days you might do two chapters. Some days you might only get through a half of a chapter because you may be giggling too much. When Fred falls down the escalator, you might feel the need to stop and send him a get-well card.

At the end of each chapter is a *Your Turn to Play*. The questions in the *Your Turn to Play* are not like the drill-and-kill exercises in most math books—forty problems all alike. Question #3 in one of the chapters is, "Could the mother have long eyelashes?"

Not like this

Right after these questions are the complete solutions. Everything is all worked out. Several mothers have e-mailed me with thoughts like, "My kid just reads the question and then reads the answer. Could you please hide the complete solutions in the back of the book or in some teacher's manual so my kid won't cheat? When they cheat, they don't learn as much."

Those moms are right when they say that just reading the question and reading the answer doesn't do much for the learning process. Just reading the question and then reading the answer is so p-a-s-s-i-v-e. It's like watching television.

Human beings learn a lot more when they are actively involved in the process. It is one thing to read about the genetics of long and short eyelashes. It is another thing to actively figure out, "Could the mother have long eyelashes?" Would you want to only read the book *Life of Fred: Eating Pizza* or would you like to engage in some practice?

THE REAL QUESTION

The real question is whether you are willing to put out the effort of reading the *Your Turn to Play* questions and answering them on your own before you look at the answer I supply.

If you are not, then my hiding the answers in the back of the book won't stop you. Besides teaching you about pre-algebra, biology, and silver wedding axes, this book can also teach you how to handle temptation—IF you want to handle it. Many ministers preach sermons about dealing with temptation in your life, but the sad truth is that many people in the congregation don't want to resist temptation in the first place.

Everybody (over the age of seventy) remembers the blond movie star Mae West and her line: "I generally avoid temptation unless I can't resist it." She is not a good role model.

If you don't want to "accidentally" see the complete solutions before you figure out the answer on your own, simply place
∗ your hand
  ∗ the corner of a pizza box
    ∗ a baseball card
      ∗ a page from the phonebook
        ∗ a handkerchief or
          ∗ a scrap of paper
over the complete solutions until you are ready to see them.

VARIOUS FONTS

There are three fonts of type that are used extensively in the book. The text of the book is written in this font. It is called Times New Roman.

When Fred is thinking, I will put his thoughts in this font.

When you, the reader, want to interject your thoughts, **you express yourself in this font.**

WORRIES THAT SOME PARENTS HAVE

This is a biology book. You can stop worrying. Fred is much purer than Mae West.

No chapters on excretion. We never mention what you do when you "go to the bathroom."

No sex stuff—although we happen to mention that Fred has a Y chromosome.

No mention of E▬▬▬▬▬▬n.∗

∗ See. I didn't mention Evolution even once.

USING CALCULATORS?

They are okay.

Wait a minute! I, your reader, have a question. Didn't you say in Life of Fred: Decimals and Percents, "Once the students get to algebra they can take their calculators out of their drawers and use them all they like"?

This is not yet algebra, and now you say it is okay to use calculators. Aren't you contradicting yourself?

No, I'm not.

Intermission
A Short Course in Logic

In the previous book, I wrote that once you get to algebra, then you can use your calculator.

If we let A = "get to algebra" and we let C = "use your calculator," then in symbolic logic, this is written as

$$A \rightarrow C. \quad \text{``}\rightarrow\text{''} \text{ means } \textit{implies}.$$

A → C is not the same thing as C → A. What C → A means is that if you can use your calculator, then you have gotten to algebra.

A → C is true. Once you get to algebra, you can use your calculator.

C → A is false. Being allowed to use your calculator doesn't mean that you have gotten to algebra.

Let's take an everyday example. Let W = my desktop computer is working. Let P = it is plugged in.

Then W → P is true. If my desktop computer is working, then it must be plugged in.

But P → W is false. If my desktop computer is plugged in, that doesn't mean it is working . . . unfortunately.

THE BRIDGES

After every seven or eight chapters, you will encounter **The Bridge**. It is all fully explained right after Chapter 7.

After the last chapter in the book, is **The Final Bridge**.

Of course, The Bridges are not just quizzes like the ones you might find in any other pre-algebra/biology book. You will learn a lot about fishing, becoming a knight, buying a dog, and things like that.

READING SPEED

Reading poetry written in Italian,
    reading the adventures of Fred, Joe, and Darlene, and
        reading mathematics and biology

      all require different reading speeds.

If you are giggling as you read, you are reading at the correct speed. If you make "silly errors" when you work on the questions in The Bridges, you might want to slow down a bit.

Some parts of the pre-algebra and the biology you might even want to read twice in order to make the ideas comfortable in your head.

There are times in life in which every word counts. This is often true in science and math.

In contrast, when reading some old-fashioned novels, I sometimes skip over whole paragraphs in which the novelist is describing in infinite detail: the lemon-yellow leaves on the stately corn stalks that were silhouetted in the amber skies provided rest for the multitudes of migrant moths as sibilant* echoes off the white-washed walls of the old fort announced that a tire of his 1936 Studebaker had developed a leak.

∗ Sibilant means hissing. You can make sibilant sounds by forcing air through a constricted passage in your mouth. Try pronouncing f, s, z, or th as in "thin."

Contents

Chapter 1	Living Things....................................	23
	the symbol >	
	adding fractions with the same denominator	
Chapter 2	What is Life?....................................	26
	union of sets	
	disjoint sets	
	finding a definition of alive	
Chapter 3	Circular Definitions.............................	30
	translating *Divina Commedia* into English	
	mixed numbers	
	improper fractions	
	multiplying and dividing fractions	
Chapter 4	Starting a Garden................................	33
	how Italian babies learn Italian	
	Victory gardens of WWII	
	multiplying by zero	
Chapter 5	Seeds and Water.................................	37
	germination of seeds	
	area of a rectangle	
	volume of a cube	
Chapter 6	Cones..	40
	ordinal numbers	
	volume of a cone: $v = \frac{1}{3}\pi r^2 h$	
	circumference and diameter of a circle	
	definition of π	
Chapter 7	Life Cycles......................................	43
	making a budget	
	2% of $500	
	four ways plants can make new plants	

The Bridge (with five tries) following Chapter 7............ 48

Chapter 8	Tooth Brushing..................................	53
	when an apostrophe is needed	
	using a common denominator to add fractions	
	why we brush our teeth	

Chapter 9	Gardening Mall Announced............................ 57
	10¢ compared with .10¢
	d = rt distance equals rate times time
	dimensional analysis
Chapter 10	At the New Mall.. 61
	20% discount
Chapter 11	Seed Money... 65
	Carolus Linnaeus classifies every plant and animal
Chapter 12	Getting Things Arranged............................. 69
	the five kingdoms
	Carl von Linné classifies every plant and animal
	kingdoms, phyla, classes, orders,
	families, genera, species
	Zea mays
Chapter 13	Blood and Your Brain................................. 72
	fainting
	which fraction is larger
Chapter 14	Conversion Factors.................................... 75
	canceling units
	subtracting decimals
Chapter 15	Plants Don't Eat Dirt.................................. 78
	Jan Baptist van Helmont and 200 pounds of dirt

The Bridge (with five tries) following Chapter 15............. 82

Chapter 16	Out of Thin Air... 87
	CO_2 supports the biology of the whole earth
	sets and subsets
Chapter 17	Photosynthesis.. 91
	chlorophyll
	sugars, starches, oils, and wood
	million, billion, trillion, and quadrillion
Chapter 18	Fred the Heterotroph................................. 95
	incisors and molars
	two uses of saliva
	digestion
	liters and meters
	autotrophs

Chapter 19	Eyes, etc................................ 100
	thermoreceptors
	chemoreceptors
	electroreceptors
	audioreceptors
	mechanoreceptors
	proprioceptors
	photoreceptors
Chapter 20	Negative Numbers........................ 103
	property liens
	worse than dead broke
Chapter 21	Eyelashes............................... 107
	two versions of the gene
	polygenic traits
	Gregor Mendel
	genotypes
	the Double R Rule
	dominant genes
	phenotypes
Chapter 22	Variation............................... 111
	squaring a number
	$x + x = 2x$ true for all numbers
	$x + 1 = x$ is never true
	recessive genes

The Bridge (with five tries) following Chapter 22............. 114

Chapter 23	Blood.................................. 119
	William Harvey
	common knowledge for 1400 years was wrong
	arteries and veins
	Marcello Malpighi finds the missing link between
	arteries and veins = the capillaries
Chapter 24	Staying Alive........................... 125
	four basic food groups according to Joe
	sugar, salt, caffeine, and saturated fat
	what can bring on hypertension
	what is hypertension
	what hypertension can cause
	blood pressure—systolic reading

Chapter 25	Solving Algebraic Equations............................ 129
	$4x + 16 + 5x = 18 + 3x + 19 + 21$
	combining like terms
	exponents
	squared, cubed, to the fourth power
Chapter 26	The Second Step in Solving Equations............... 134
	volume of a cylinder $V = \pi r^2 h$
	a zillion
	why I became a math major
	right triangles
	$y^3 = 27$
	$x^x = 4$
Chapter 27	Words to Equations...................................... 141
	three suggestions for turning
	word problems into equations
	drawing diagrams
Chapter 28	Breathing... 146
	the triple treatment: warmed, moistened, and cleaned
	why fish don't breathe in and out
	basal metabolic rate
Chapter 29	Still Breathing... 150
	fogging a mirror
	the aorta
	the windpipe—the trachea

The Bridge (with five tries) following Chapter 29............. 154

Chapter 30	How Oxygen Is Carried in the Blood................. 159
	hemoglobin
	simile and metaphor
	chlorophyll vs. hemoglobin vs. hemocyanin
	glucose
	crustaceans
Chapter 31	Large Numbers... 163
	a quadrillion dollars to spend
	Avogadro's number
	602,213,670,000,000,000,000,000
	atomic weights
	five meanings of the word *mole*

Chapter 32	Stoichiometry... 169
	balancing chemical equations
	coefficients
	two giant stoichiometry hints
Chapter 33	Small Numbers... 173
	sucrose
	how many tons does one sucrose molecule weigh?
	drawing mazes
	the whole numbers {0, 1, 2, 3, . . . }
Chapter 34	Shortcuts.. 177
	sweet and greasy
	why lions have more free time than antelopes
	pretending that the future doesn't exist
	smoking, junk foods, cheating, and lying
Chapter 35	Reducing Fractions..................................... 181
	when numbers are divisible by 5
	when numbers are divisible by 2
	when numbers are divisible by 3
	the four General Rules for fractions
Chapter 36	Division by Zero.. 185
	why it is not permitted
	arguments by contradiction—indirect proofs
	cross multiplying

The Bridge (with five tries) following Chapter 36.............. 189

Chapter 37	Bones 198
	solipsism
	cranium, sternum, femur, and patella
	cartilage
	why you don't give food or water to someone with a possible bone fracture
Chapter 38	Better Bones... 203
	calcium needs
	osteoporosis
	4 lbs. of tofu per day
	if the love of pizza were a dominant gene

Chapter 39	Integumentary System.	207

     farther and *further* in informal English,
       general English, and formal English
     neurons

Chapter 40	Epidermis.	210

     mitotic division
     why we like dead, flattened out epidermal cells
     two uses of your epidermis
     rifle bullets and rhino skin
     doubling and re-doubling your epidermis

Chapter 41	Dermis.	216

     hair
     sweat glands
     oil glands
     your dermis—official excuse for watching
       mindless television

Chapter 42	Genes.	222

     mutations—regular, rare, and random
     meiosis
     mitosis
     locus of a gene

Chapter 43	Six Words.	229

     chromosomes
     in 1922, we had 24 pairs
     in 1955, we had 23 pairs
     we are diploid
     triploid, tetraploid, hexaploid, and octaploid
     your aloneness when you say 23, and everyone else
       says 24
     Down syndrome

Chapter 44	Five Words..	234

     the names of each of the 23 pairs of chromosomes
     what to do if you find out that your twenty-third pair
       is mismatched—one normal sized and the
       other puny.

Chapter 45	Five Words: Locus, Site, DNA, Gene, Allele.	238

     Watson and Crick and the structure of DNA
     nucleotides
     the Human Genome Project
     we are 99.99% alike

Chapter 46   Entre Nous.................................. 241
                      color blindness
                      site
                      point mutation
                      alleles
                      nine hairy complications of biology
                      junk DNA
                      pleiotropic genes
                      incomplete dominance
                      codominant alleles
                      how to affect your phenotype after you are born

The Final Bridge (with five tries) 248

Answers to **The Bridges**............................. 264

Index... 282

Chapter One
Living Things

It was just before dawn. Fred was
        in his sleeping bag,
                under his desk,
                      in his office,
                            in the math building
                                  at KITTENS University
                                        in Kansas.
And he was asleep. Dreaming.

    You might think that someone who taught arithmetic, beginning algebra, advanced algebra, geometry, trigonometry, calculus, statistics, and linear algebra would have dreams filled with

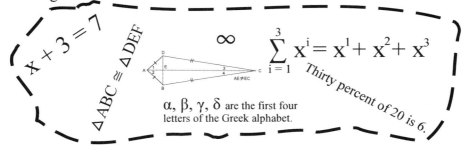

α, β, γ, δ are the first four letters of the Greek alphabet.

    And some days, those were his dreams. But not this morning. Our five-and-a-half-year-old professor of math dreamed about a cow having a

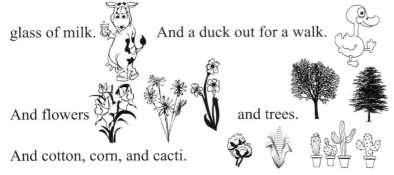

glass of milk.   And a duck out for a walk.

And flowers   and trees.

And cotton, corn, and cacti.

Chapter One   Living Things

Have you ever had a dream and wondered what it meant? This wasn't Fred's typical math dream. If he had dreamed,

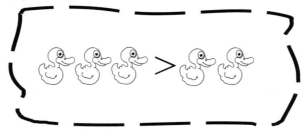

then that would have been easy to understand. Three ducks are greater than two ducks. Just like 8 > 6.

Or if he had dreamed of adding fractions with a common duck—I mean with a common denominator—that wouldn't have puzzled Fred.

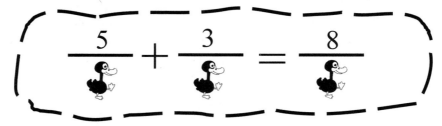

Cows and ducks, flowers and trees, cotton, corn, and cacti—what's this dream all about? Then he dreamed of this dog with the answer:

On the next page it's *Your Turn to Play* . You learn by reading, and you also learn by doing. This is your turn to *do*. Take out a piece of paper and write your answers down before you look at the solutions I have provided.

Chapter One   Living Things

Did you know that teachers used to be students? Why are they now smarter than the students they teach? One reason is that they *do* the stuff. They stand at the blackboard, and they write and talk.

Please don't just read the questions and then read the answers. Use a piece of paper and a pencil.

Please.

Your Turn to Play

1. When scientists study the stars and the moon, it's called astronomy. When they study matter, energy, motion, and force, they are studying physics. What do they call the study of living things? (Hint: Look at the title of this book.)

2. Which of these are true?

$7 > 4$

$\frac{44}{11} > 5$

$\pi > 3$

3. Let's add some fractions with a common ~~duck~~ denominator.

$\frac{4}{13} + \frac{7}{13} = ?$

. . . . COMPLETE SOLUTIONS . . . .

1. If you guessed "Pre-algebra," you only get half credit. The study of living things is called biology.

2. $7 > 4$ is true. Seven is greater than four.

    $\frac{44}{11} > 5$ is false. Four is not greater than five.

    $\pi > 3$ ($\pi \approx 3.14159265358979323846264338327950 28$ where $\approx$ means *approximately equal to*.)

Since $3.14159265358979323846264338327950 28 > 3$, we know that $\pi > 3$ is true.

Many students memorize that $\pi \approx 3.14$ or that $\pi \approx 3\frac{1}{7}$.

3. $\frac{4}{13} + \frac{7}{13} = \frac{11}{13}$ not $\frac{11}{26}$

Chapter Two
What is Life?

Fred was very glad that the dog told him what his dream was about. The dog had told him, "it's about life." If you were a biology teacher, you could brag, "I teach about life."*

As Fred's dream continued, he saw everything in the physical universe divided into two sets:

Combined together, these two sets encompassed every physical thing in the universe. Every physical thing.** Another way of saying that is that the **union** of the two sets is equal to every physical thing in the

---

★ Of course, the debate goes on between biology teachers and math teachers. The math teachers say to the biology teachers, "Why—if you teach about life—are there so many dead things, like dead frogs, in your biology labs?"

But the biology and the math teachers agree that the cooking classes (culinary arts) are a lot worse. They have lots of dead things in their kitchens.

But biologists never have to eat their dead stuff.

★★ Every *physical* thing. There are lots of things that are real that are not physical. Love is real. The number 7 is real. Time is real.

universe. He also noticed that those two sets were **disjoint**—something was either not living or it was living. Of course, when your pet worm dies, it gets transferred from the *alive* set to the *not-alive* set. But at any instant in time, you are either in one set or the other, but not in both.

    Fred's eyes popped open. It was 5 a.m. He wiggled out of his sleeping bag and rolled it up. Actually, to tell the truth, it wasn't really a sleeping bag, but just a big pillowcase. When you are only three feet tall, you don't need a six-foot sleeping bag. He put the rolled-up pillowcase into a desk drawer.

    He climbed onto his desk chair, which had three telephone books on it so that he could be tall enough.

    If someday I write a biology book he thought to himself *I'll start it with a definition. In geometry class, I defined what each of the geometry figures was . . .*

I should do the same thing in biology.
Fred took out his clipboard and uncapped his fountain pen and began . . .

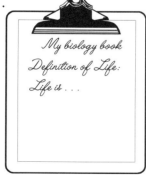

    He was stuck. He sat there for five minutes and couldn't finish the definition of the word *life*.

    I know that I am alive he thought. *I know that my fountain pen is not alive. But how do I write a definition so that someone using the definition can tell whether something is alive?*

Chapter Two   What is Life?

Your Turn to Play

1. What's wrong with: *If something is alive then it can smile.*
2. What's wrong with: *If something is alive then it can move.*
3. What's wrong with: *If something is alive if it can grow.*
4. What's wrong with: *Something is alive if when it is started, it will continue on its own.*
5. What's wrong with this definition: *If something is alive then it can have babies.*
6. What's wrong with this definition: *Something is alive if it or any of its kind can reproduce.*
7. What's wrong with this definition: *Something is alive if it can respond to a change in the environment.*✶
8. Is it true that it is difficult to write a definition of what it is to be alive? Check one box: ☐Yes ☐Yes.
9. Here is one set: {Texas, Maine, Arizona}. Here is another set: {Washington, Arizona, South Carolina, North Carolina}. Are these sets disjoint?
10. A culinary arts student responded to the footnote of two pages ago. She said, "Of course we all eat dead stuff. Carrots and steak don't say 'ouch!' when you bite them. We're not tigers or sharks that go around munching on living things."

    I went to the fridge and pulled out a container of food that bragged that it contained living organisms. I popped off the lid and had a big spoonful of it. What was I eating?
11. Consider the set of all people whose *first* name begins with J and the set of all people whose *last* name begins with M. Are these two sets disjoint?

✶ I was going to write, *Something is alive if it can react to a stimulus.* But that would have been dumb. Why? Because a *stimulus* is, by definition, that which causes a reaction. The minute that you declare something to be a stimulus, you have already said that a reaction will occur.

Chapter Two   What is Life?

·······COMPLETE SOLUTIONS·······

1. Ducks can't smile. Their bills don't bend. (Your answer may be different than mine.)

2. Moss and seeds don't move, but they are alive.

3. Balloons, bank accounts, and the national debt can grow, and they are not alive.

4. Walk into a giant field of tall dry wheat and light a fire. The flames will continue on their own, but the fire is not alive. (Note: You might not be alive after this experiment.)

5. If your cat has been neutered ("fixed"), it can't have babies. But your cat is still alive. Male mules are infertile.*

6. This is a more difficult question. Credit card balances, if left alone, seem to multiply. On a more serious note, there are computer programs that can reproduce themselves.

7. A thermometer responds to a change in the environment. So does a thermostat.

8. The correct answer is ☒Yes.

9. The sets {Texas, Maine, Arizona} and {Washington, Arizona, South Carolina, North Carolina} are not disjoint because they both contain Arizona.

10. The container has on the label that it contains live acidophilus and bifidus—millions of them. Yogurt.

11. My favorite singer in the movies is Jeanette MacDonald. The two sets are not disjoint.

        James Madison, United States president
        James Monroe, United States president
        Joe Montana, football player
        James Mason, actor
        Johnny Mathis, singer
        Justin Martyr, Bishop of Antioch in 150 AD
        Judy Martin, professional wrestler and former WWF Women's Tag Team Champion

---

✶ Here are the details for those who already know some biology. Donkeys have 62 chromosomes. Horses have 64 chromosomes. They are different species. The offspring of a male donkey and a female horse is a mule.
    Most, not all, female mules are infertile.

Chapter Three
Circular Definitions

Fred shrugged his shoulders. He reached for his dictionary and looked up the word *life*. It read: LIFE: THAT WHICH IS ALIVE. Then he looked up *alive* and found: ALIVE: THOSE THINGS THAT ARE LIVING. In desperation, he then looked up *living*: LIVING: A PROPERTY OF LIFE.

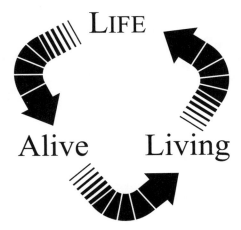

Fred was caught in a circular definition.

short story

Two years ago when Fred was three, he was looking through Harold Bloom's *The Western Canon* to see a list of the great works of world literature. He was surprised that there were no titles like Prof. Eldwood's *Guide to Pizza*, but he did find the central poem of Italian literature: Dante's *Divine Comedy*.

Fred was in the mood for something humorous, so he headed off to the bookstore and bought a copy of the *Divine Comedy*.* When Fred got

---

* Fred didn't know that there are two meanings of *comedy*. One meaning is the ha-ha-ha portrayal of amusing events. The other meaning of *comedy* are those works that have a happy ending. Since the first third of Dante's *Divine Comedy* takes its readers down through all the circles of Hell, right down to the innermost circle in which Satan is found half-buried in ice, it would be a good guess that Dante wasn't writing a tickle-and-giggle comedy.

Chapter Three Circular Definitions

back to his office, he realized that the bookstore operator had sold him *Divina Commedia*, which is the *Divine Comedy* in the original Italian.

Great! thought Fred. This will give me a chance to learn Italian. Fred opened the book and began to read:

> *Nel mezzo del cammin di nostra vita*
> *mi ritrovai per una selva oscura*
> *ché la diritta via era smarrita.*
> *Ahi quanto a dir qual era è cosa dura*
> *esta selva selvaggia e aspra e forte*
> *che nel pensier rinova la paura!*

He liked the exclamation point at the end of the sixth line. He thought That means something exciting is happening. But I don't even know what the first word *nel* means. Fred ran back to the bookstore and asked for an Italian dictionary. The bookstore owner handed him *Stanthony's Italian Dictionary*. It weighed eight pounds.

Fred hauled it back to his office, hopped on his chair with the three telephone books and opened *Stanthony's Italian Dictionary* to the word *nel*. The dictionary defined *nel* as *dentro*. What in the world is *dentro*? Fred thought.

He looked up *dentro* and found *nel*. Then he realized that he had bought a real Italian dictionary, the kind that Italians use. It wasn't an Italian-English translation dictionary. This was Fred's first introduction to circular definitions.

———end of short story———

In case you are interested, the first line *Nel mezzo del cammin di nostra vita* means *Midway upon the journey of our life*.

Your Turn to Play

1. Fred thought about those words from the *Divine Comedy: Midway upon the journey of our life*. If he were half way through his life right now, how old will he be at the end of his life? (As mentioned on the first page of Chapter 1, Fred is now 5½ years old.)

Chapter Three   Circular Definitions

2. If Fred were going to live to the age of 80, and if he were half way through his life, how old would he be now?

3. $3\frac{1}{8} \div 3\frac{1}{5}$

.... COMPLETE SOLUTIONS ....

1. If 5½ is half of the total length of his life, we will have to double 5½ to find the total length.

$$5½ \times 2 = \frac{11}{2} \times \frac{2}{1} = \frac{22}{2} = 11$$

Looking Back

5½ is a **mixed number**. In order to multiply a mixed number, you first need to change it into an **improper fraction**.
To change $5\frac{1}{2}$ into an improper fraction, you say in your head: *2 times 5 . . . plus 1.*

Looking Back

To multiply fractions, you multiply

top times top

and

bottom times bottom.

2. There are several ways you might have solved this.

Way #1: $\frac{1}{2}$ of $80 = \frac{1}{2} \times 80 = \frac{1}{2} \times \frac{80}{1} = \frac{80}{2} = 40$

Way #2: $80 \div 2 = \frac{80}{1} \div \frac{2}{1} = \frac{80}{1} \times \frac{1}{2} = \frac{80}{2} = 40$

Looking Back

To divide fractions, you invert the fraction on the right and multiply.

Way #3: $2\overline{)80}^{\,40}$

3. $3\frac{1}{8} \div 3\frac{1}{5} = \frac{25}{8} \div \frac{16}{5} = \frac{25}{8} \times \frac{5}{16} = \frac{125}{128}$

Chapter Four
Starting a Garden

Fred wondered how Italian babies ever learn Italian. When they are born, they don't know any Italian. For example, they don't know what the word *nel* means. If they ask their mom, "What does *nel* mean?" she might say, "*Nel* means *dentro*." And if they ask her what *dentro* means, she would say that it means *nel*.

When you learn a second language, things are different. You can always ask, "What does *nel* mean in English?"

But how do Italian babies learn Italian and Greek babies learn Greek, and German babies learn German?

The answer is that mothers do a lot of pointing.

dog

Just then, Fred heard the morning newspaper hit the door of his office. Fred closed *Divina Commedia* and his Italian dictionary. He hopped off his chair, went and got the paper, and headed back to his chair to read . . .

The KITTEN Caboodle

The Official Campus Newspaper of KITTENS University   Friday 6:10 a.m. Edition 10¢

Everyone's Planting a Victory Garden!

KANSAS: Our University president has just discovered some Victory Garden posters in his closet. He has ordered copies of them made and plastered all over the campus.

"Victory Gardens sounds cool," he announced this morning at a press conference. The reporters who showed up for the 5 a.m. meeting couldn't understand what the excitement was all about.

"We must plant, plant, plant!" were the president's last words before he ended the meeting (and went back to bed).

Our University president has stated, "We must have victory."

One reporter asked him, "Victory over what?"

The president wasn't sure.

Chapter Four   Starting a Garden

Fred heard someone at the door to his office. He hopped off his chair again and opened the door. The janitor had just pasted a giant poster on his front door. He looked down the hallway and every door was covered with a copy of this poster.

The poster covered Fred's room number and the schedule of classes he taught.

on Fred's front door

> Fred Gauss
> —room 314—
> 8–9   Beginning Algebra
> 9–10 Advanced Algebra
> 10–11 Geometry
> 11–noon Trigonometry
> noon–1 Calculus
> 1–2   Statistics
> 2–3   Linear Algebra
> 3–3:05 Break
> 3:05–5 Seminar in Biology, Economics,
>   Physics, Set Theory, Topology,
>   and Metamathematics.

Victory gardens? Fred thought to himself. *That poster is from World War II.* Didn't someone tell the president that the war ended in 1945?*

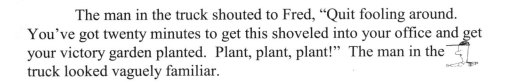

Fred rubbed his eyes. He couldn't believe what he saw coming down the hallway. *That's not possible!* It was a giant dump truck. *I'm on the third floor of this building. How did it get up here?*

The dump truck dumped. Fred was up to his neck in dirt.

The man in the truck shouted to Fred, "Quit fooling around. You've got twenty minutes to get this shoveled into your office and get your victory garden planted. Plant, plant, plant!" The man in the truck looked vaguely familiar.

★ Back in the 1940s everyone was encouraged to plant vegetable and fruit gardens in their backyards or in vacant lots. Canned goods were being rationed. Victory gardens allowed the canned peas, carrots, etc. to be sent to the troops.
   There were victory gardens in the United States and in Germany.

Chapter Four   Starting a Garden

"And hand these out to your students." The driver placed a stack of victory garden posters on top of Fred's head and drove down the hall.

Fred's arms were trapped in the dirt. He looked like a little table with literature on it. His nose itched and he couldn't scratch it.

It was 6:20 and his first class was at 8. Students came by and took posters off the "table." Fred was too embarrassed to call for help. He was afraid that the photographers would come, and he would end up in the newspaper.

In ten minutes, his fears came true.

The KITTEN Caboodle

The Official Campus Newspaper of KITTENS University   Friday 6:30 a.m. Edition 10¢

exclusive

Prof. Fred Plays in the Dirt and Gets Stuck

KANSAS: All of the campus is laughing. Five-and-a-half-year-old professor Fred Gauss was apparently playing in some victory garden dirt, and he got stuck. Everyone has come by to look at him. They all wondered how a kid could be so careless.

One mother of three commented, "My kids have played in the dirt for years, and they never have gotten stuck. This Gauss guy may know mathematics, but he doesn't know dirt."

Big Mistake by Kid Prof

One student asked Fred how he would get to his eight o'clock class this morning. "It looks like your arms are trapped," he told Fred.

The head of KITTENS University Human Resources division came by and asked Fred if they should hire a long-term substitute to replace him.

Sally, 2, asked, "Mommy, why doesn't anyone help Freddie get out of the hole?"

Sally's mother told her, "It's not polite to stare at people with handicaps" and pulled her away.

advertisement

Glenda's Garden Goodies

Victory Garden Special!

Celery Seeds
15¢
(0.37 gram packet)

Chapter Four   Starting a Garden

Your Turn to Play

1. A packet of Glenda's celery seeds is 15¢. How much would 6 packets cost?

2. How much would a hundred packets cost?

3. One packet of celery seeds weighs 0.37 grams. How much would a thousand packets weigh?

4. If each packet had 58 seeds, and if each seed could produce a celery plant weighing 1.62 pounds, and if each celery plant contained 7 stalks of celery, and if each stalk contained zero calories, how many calories could you get if you planted 9862 packets? (Hint: multiply.)

. . . . COMPLETE SOLUTIONS . . . .

1. 6 × 15¢ = 90¢

This answer could also be written as $0.90.

2. 100 × 15¢ = 1500¢

This answer could also be written as 1500.¢ or as $15.00.

     If the number doesn't have a decimal point, multiplying by a 100 is done by just adding two zeros.

     For example: 100 × 8234 = 823400

  A second way: ① 15 is the same as 15. since you can always add a decimal point to the right of a number that doesn't have a decimal point.

           ② 15. is the same as 15.00

           ③ To multiply by a hundred, move the decimal point two places to the right. 15.00 ➡ 1500.

3. 0.37 is the same as 0.370, and when we multiply by 1000, we move the decimal three places to the right and get 370. grams.

     370. grams could also be written as 370 grams.

4. 58 × 1.62 × 7 × 0 × 9862 is the world's easiest multiplication problem. If one of the factors in a multiplication is zero, then the answer will be zero. 0 × anything = 0.

Chapter Five
Seeds and Water

Fred was encased in this giant block of dirt. It was as wide as the hallway, which is 8 feet. It was 7 feet long.

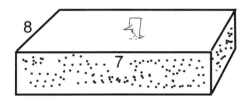

Will I be here for the rest of my life? Fred thought. My students could come here and sit on the dirt, and I could teach them. Or this could be a playground, and little kids could come and play in the dirt. I would watch and make sure they didn't get hurt.

The top of the dirt is a nice rectangle. For rectangles, length times width equals the area. So we could call this the Playground with 56 Square Feet.

No kids came. Except for Sally's brother Jimmy. Jimmy looked at Fred for a moment and asked, "Are you a scarecrow?" Jimmy's mother came and said to him, "I've been looking all over for you." With Sally in one hand and Jimmy in the other, she headed down the hallway. Sally turned and winked at Fred.

It was quiet. Fred was buried in his 56 square feet of dirt. If I had some seeds in my mouth, I could spit them out and plant a victory garden. Then the dirt wouldn't go to waste.

Seeds **germinate*** when it's warm and they get wet. It was warm in the hallway, but Fred wasn't sure he could spit enough to keep the seeds watered.

★ Germinate = start to grow

Chapter Five   Seeds and Water

If I had a free hand and a shovel, I could dig my way out. But only my head is free. Then the solution came to Fred. He had the perfect shovel.

In ten minutes he dug himself out. Then he walked down the hallway to the restroom to wash the dirt off his nose.

When he got back, the dirt in the hallway was all gone! *Disappearing dirt,** Fred thought.

The janitor was standing there with a shovel and informed Fred, "I got rid of this mess in the hallway. Enjoy your garden."

Fred didn't know what he meant by "Enjoy your garden." He opened the door to his office—the one with the 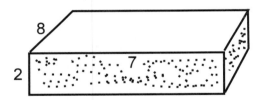 poster on it. He now knew where the dirt went.

Your Turn to Play

1. Suppose the giant block of dirt in the hallway were 8 feet wide and $7\frac{1}{2}$ feet long instead of 7 feet. What would be the area of the top of the block?

2. Suppose the block of dirt in the hallway was 2 feet deep.

The volume of a block is length times width times depth. In algebra they are going to write this in symbols as: v = lwd. Since the letter "l" looks so much like the number "1," they might write that formula in cursive: $v = lwd$.

What is the volume of a block that is 8 feet wide, 7 feet long, and 2 feet deep?

★ *Disappearing dirt* is an example of **alliteration**. That's a kind of rhyme where the first sounds of the words are alike. *Sing several silly songs. Who hates heavy hats? Wicked witches wilt when watered.*

Chapter Five   Seeds and Water

3. True or False: $2\pi < 6$ $\dfrac{25}{5} < 2\pi$ $25\% < \dfrac{1}{4}$

4. $1\dfrac{1}{8} \div 1\dfrac{1}{4}$

. . . . COMPLETE SOLUTIONS . . . .

1. $8 \times 7\dfrac{1}{2} = \dfrac{8}{1} \times \dfrac{15}{2} = \dfrac{120}{2} = 60$ square feet

Or this could have been done with canceling:

$8 \times 7\dfrac{1}{2} = \dfrac{8}{1} \times \dfrac{15}{2} = \dfrac{\cancel{8}^{4}}{1} \times \dfrac{15}{\cancel{2}_{1}} = \dfrac{60}{1} = 60$ square feet

Or it could have been done with decimals:

$8 \times 7\dfrac{1}{2} = 8 \times 7.5 = \begin{array}{r} 7.5 \\ \times\ 8 \\ \hline 60.0 \end{array}$ square feet

2. $8 \times 7 \times 2 = 56 \times 2 = 112$ cubic feet

3. Since π is greater than 3, $2\pi < 6$ is false and $\dfrac{25}{5} < 2\pi$ is true. $25\% < \dfrac{1}{4}$ is false because 25% equals $\dfrac{1}{4}$

4. $1\dfrac{1}{8} \div 1\dfrac{1}{4} = \dfrac{9}{8} \div \dfrac{5}{4} = \dfrac{9}{8} \times \dfrac{4}{5} = \dfrac{9}{\cancel{8}_{2}} \times \dfrac{\cancel{4}^{1}}{5} = \dfrac{9}{10}$

UNITS OF MEASURE

When you measure distance, your answer may be something like 7 feet (or 7 miles or 7 inches).

In question 1, we were measuring area. Our answer was given in *square* feet. Imagine a chessboard (or checkerboard) filled with one-inch squares. The area of that would be 64 square inches.

In question 2, we were measuring volume. Our answer was given in *cubic* feet. Imagine that the dirt was divided up into little cubes with edges that are one foot long. The volume is expressed as 112 cubic feet.

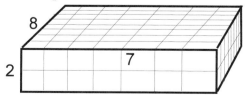

Chapter Six
Cones

The first thing that Fred noticed when he opened the door to his office was that the janitor had shoveled all the dirt from the hallway into his office.

The second thing that he noticed was that the dirt was piled up in the shape of a cone. The third thing he noticed was that his desk was buried under all the dirt. *First, second,* and *third* are **ordinal numbers**. They tell you what *order* things are in.

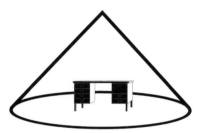
Fred's desk was inside a giant cone of dirt

The radius of that cone of dirt was six feet. Its height was seven feet.

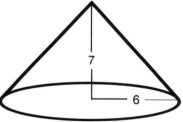

The **volume of a cone** is $v = \frac{1}{3}\pi r^2 h$. This is going to take some explaining.

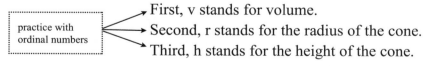

First, v stands for volume.
Second, r stands for the radius of the cone.
Third, h stands for the height of the cone.

[practice with ordinal numbers]

Chapter Six   Cones

Fourth, pi (which is written as π and pronounced "pie") is a number. π is a little bit bigger than 3. π > 3. It is a little bit bigger than 3.14. π > 3.14. It is a little bit smaller than $3\frac{1}{7}$. π < $3\frac{1}{7}$.

If you were to take a string and make it into the shape of a circle, the length of the string would be the **circumference** of the circle.

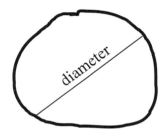

If you divide the circumference by the **diameter** of the circle (the distance across the circle) you get π. That's the definition of π.

π is a little bit bigger than 3.1415 and a little smaller than 3.1416. 3.1415 < π < 3.1416.

If you had created the world, you might have made some things different:
* You would never have to go to the dentist.
* It would always be springtime.
* Every Thursday would be Christmas.
* Every Sunday would be Easter.
* And π would be equal to 3 exactly.

For rough work—use 3 for π.
For good work—use $3\frac{1}{7}$ for π.
For better work—use 3.14 for π.
For even better work—use 3.1416 for π.
Fred uses 3.14159265358979323846264338327950288 for π.

Fifth, r² means "r times r." The formula for the volume of a cone (v = ⅓πr²h) could have been written as v = ⅓ πrrh.
r² is pronounced "r squared."
x^{42} ("x to the forty-second power") is
xxxxxxxxxxxxxxxxxxxxxxxxxxxxxxxxxxxxxxxxxx.

41

Chapter Six Cones

Your Turn to Play

1. 3^4 means 3 times 3 times 3 times 3, which is 81. Which is larger 2^5 or 5^2 ?

2. Simplify $1^{9083276409262390}$.

3. The radius of that cone of dirt was six feet. Its height was seven feet. Using 3 for π, what is the volume of that cone?

4. (Continuing the previous question) if we use $3\frac{1}{7}$ for π, what is the volume of the cone of dirt?

5. Suppose the edges of a cube of sugar are $\frac{2}{3}$ inches. What is the volume of that cube?

6. $(\frac{3}{4})^3 = ?$

7. Express $\$10^6$ in words.

·······COMPLETE SOLUTIONS·······

1. 2^5 means 2 times 2 times 2 times 2 times 2, which is 32.
$5^2 = 5 \times 5$, which is 25. So $2^5 > 5^2$.

2. $1^{9083276409262390}$ means 1 times itself 9083276409262390 times. If I work it out $1 \times \ldots \times 1$ equals 1.

3. $v = \frac{1}{3} \pi r^2 h = \frac{1}{3} \times 3 \times 6^2 \times 7 = 252$ cubic feet.

4. $v = \frac{1}{3} \pi r^2 h = \frac{1}{3} \times 3\frac{1}{7} \times 6^2 \times 7 = \frac{1}{3} \times \frac{22}{7} \times \frac{36}{1} \times \frac{7}{1} = \frac{1}{3} \times \frac{22}{7} \times \frac{36}{1} \times \frac{7}{1} = 264$ cubic feet.

5. $\frac{2}{3} \times \frac{2}{3} \times \frac{2}{3} = \frac{8}{27}$ cubic inches

6. $(\frac{3}{4})^3 = \frac{3}{4} \times \frac{3}{4} \times \frac{3}{4} = \frac{27}{64}$

7. $10^6 = 10 \times 10 \times 10 \times 10 \times 10 \times 10 = 1{,}000{,}000$
So $\$10^6$ is a million dollars.

Chapter Seven
Life Cycles

Fred had the dirt. He figured that that was the hard part. Next he needed the seeds for his victory garden. Seeds Fred thought to himself are a lot cheaper than buying the plants and sticking them in the dirt. Fred's salary at KITTENS University was $500/month for the nine hours he taught each day. That was a lot of money for a five-and-a-half-year-old professor, but he didn't want to spend it foolishly.

> Fred's Budget
> 10% for Sunday School offering
> 30% for books
> 1% for clothes
> 2% for buying food at the vending
>   machines down the hallway
> The rest for savings for old age

He didn't need to spend anything for housing. He slept under his office desk each night. (This may need to change if his desk remains buried under a pile of dirt.)

The money for the seeds can come out of the 2% I spend for buying food. Fred did a quick calculation:

$$2\% \text{ of } \$500$$
$$2\% \times \$500 \quad \text{("of" often means multiply)}$$
$$0.02 \times 500.$$

```
    500
 × 0.02
  10.00      $10/month for the food budget
```

Fred had visions of his seven-foot high cone of dirt covered with artichoke, asparagus, avocado, beets, bok choy, broccoli, brussels sprouts, cabbage, carrots, cauliflower, celery, chard, chile peppers, chives, collards, corn, cucumbers, dandelion greens, eggplant, endive, fennel, garlic, green beans, iceberg lettuce, Jerusalem artichoke, jicama, kelp, leeks,

Chapter Seven　　Life Cycles

mushrooms, mustard greens, okra, onions, parsnips, peas, potatoes, pumpkin, radishes, red leaf lettuce, romaine lettuce, rutabagas, scallions, snow peas, spinach, squash, strawberries, sweet peppers, sweet potatoes, tomatoes, turnips, yams, and zucchini.

Fred had read about **life cycles** in Prof. Eldwood's book *A Chicken Is the Way an Egg Makes Another Egg,* 1844.

You start with
adult chickens

You get eggs

You get baby chicks

They grow up and
you have more adult chickens

You start
with adults

You get babies
(people don't
lay eggs!)

They become kids

They grow up and
you have more adults

Chapter Seven   Life Cycles

You start with adult plants

You get seeds

Seeds germinate

They grow up and you get more adult plants

When Fred read Prof. Eldwood's book, he noticed several errors in it. In the life cycle of chickens, he had a drawing of eggs in a carton, like the kind you buy at the grocery store. For several reasons those eggs aren't going to hatch. One reason is that they have been refrigerated instead of being kept warm by having the mother hen sit on them.

Also, please don't expect to hatch eggs that have been fried.

The second error that Fred found was in the life cycle of plants. To go from one adult plant to other adult plants you don't always have to use seeds.

Strawberries are a good example. They send out "runners" which create new strawberry plants.

45

Chapter Seven   Life Cycles

And one daffodil bulb will turn into two by dividing underground. And some plants, like roses, can reproduce by cloning. You take a stem cutting and put it into the ground and keep it wet.*

> **Intermission**
>
> The *Your Turn to Plays* at the end of each chapter are your chance to practice. Take out a sheet of paper and actually DO the problems BEFORE looking at the answers.
>
> If you cheat and just look at the answers, you won't learn as much, and the Bridges will be a lot harder.

Your Turn to Play

1. Let's look at Fred's budget. He makes $500/month. He budgeted 10% for Sunday School offering, 30% for books, and 1% for clothes. How much money is in each of these three categories?
2. If he allocated 10% for Sunday School, 30% for books, 1% for clothes, and 2% for food, what percent was left for his "savings for old age"?
3. A large packet of flower seeds contains 25,000 seeds. If 3% of them germinate, how many plants will you have?

........**COMPLETE SOLUTIONS**.......

1. 10% of $500 = 0.10 × 500 = $50.
   30% of $500 = 0.30 × 500 = $150.
   1% of $500 = 0.01 × 500 = $5
2. 10% + 30% + 1% + 2% = 43%.   100% − 43% = 57%
3. 3% of 25,000 = 0.03 × 25,000. = 750 plants

✶ This is fairly tricky to do. There are gardening books that tell you all the secrets. They tell you how to make the cutting, how to use certain powders to make the roots grow better, how to choose the right kind of soil.

I have tried to make clones of roses and have always failed. I would forget to keep the soil moist.

For those of you who have read *Life of Fred: Fractions* or *Life of Fred: Decimals and Percents*, you know what this means. You can take out a piece of paper and turn to the next page right now.

For the rest of you . . .

We are at **The Bridge**.

After every seven or eight chapters, we give you the chance to show that you haven't forgotten what you have learned. **The Bridge** consists of ten questions from the beginning of the book up to the present moment.

It also gives you the chance to show that you actually worked the problems in the *Your Turn to Play*.

If you get 90% or more right, you have crossed the Bridge and have earned the right to go on to the next chapter. You are permitted to look back at earlier material in the book while you take this quiz.

After you have finished all ten problems, you and your parent (or guardian or teacher or jailer or superintendent or whoever is your supervisor) may compare your answers with those in the back of this book to see if you have crossed the Bridge.

Good luck.

The Bridge
from Chapters 1 to Chapter 7

first try

> Goal: Get 9 or more right and you cross the bridge.

1. A **cardinal number** counts the number of elements in a set. What is the cardinal number associated with {☙, 82, ✈, ✳}?

2. Are {☙, 82, ✈, ✳} and {41, ✳, α, β, γ} disjoint?

3. $\frac{9}{17} + \frac{7}{17}$

4. Complete this sentence: *The number 3.7 can't be a cardinal number because. . . .*

5. Joe and Darlene are students in one of Fred's math classes. Joe isn't paying attention 97% of the time. What percent of the time is he paying attention?

6. If a cube is 9 feet long, 6 feet wide, and 1.1 feet deep, what is its volume?

7. $10^3 = ?$

8. Darlene thinks about Joe 25% of the 16 hours she is awake each day. How many hours per day does she think about him?

9. What is the volume of this cone? (Use 3 for π.)

10. When Fred gave a test in his math class, Joe came in 82nd. Is this a cardinal or an ordinal number?

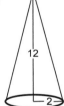

After you have written the answers to all ten problems, and after you have checked your work as much as you want to, it will be time to find out whether you have crossed the bridge. If you got nine or ten correct, you are ready to go on to Chapter 8.

If not, then you probably have been reading too fast. (Remember the story of the tortoise and the hare. The speedy one is not always the one who finishes first.) You may earn the right to another try (on the next page) by first correcting all the errors you made.

The Bridge
from Chapters 1 to Chapter 7

second try

1. $7\frac{1}{8}$ is a mixed number. Change it into an improper fraction.

2. If Fred were going to live to the age of 80, and he is one-sixteenth of the way through his life right now, how old is he?

3. Are the set of all squares and the set of all triangles disjoint?

4. Simplify as much as possible: $5^3 \times 1^{39867} \times 0^6$.

5. True or False: $\frac{88}{12} < 7.3$

6. Joe dreams about going fishing. One fishing lure cost $1.29. How much would a hundred lures cost?

7. Darlene gave Joe a sinker for his fishing gear. It was a piece of lead in the shape of a cone. It was 2 inches tall and had a radius of 0.8 inches. What was its volume? (Use 3 for π.)

8. Joe went to the bait store to buy fishing worms. He bought 1800 of them. (He counted them.) Five percent of them were dead. How many were dead?

9. $\frac{11}{103} + \frac{11}{103} = ?$

10. $\frac{2}{3} \times \frac{2}{3} = ?$ This is the same as $\left(\frac{2}{3}\right)^2$.

If you got nine or ten correct, you are ready to go on to Chapter 8. Otherwise, correct your errors, and you have earned the right to a third try.

The Bridge
from Chapters 1 to Chapter 7

third try

1. The set of things that Joe is interested in is {eating, napping, fishing}. The set of things that Darlene is interested in is {her nails, her hair, what's on TV, marrying Joe}. Are those sets disjoint?

2. 4¾ × 1⅓ = ? Simplify your answer as much as possible. Do not leave it as an improper fraction.

3. 4¾ ÷ 1⅓ = ? Simplify your answer as much as possible. Do not leave it as an improper fraction.

4. April has 30 days. On his calendar, Joe circled 30% of them. When Darlene asked Joe why he did that, Joe told her that those were his fishing days. How many days had Joe circled?

5. Darlene bought two fishing lures for Joe. One weighed $\frac{4}{9}$ of an ounce and the other weighed $\frac{1}{9}$ of an ounce. What was the total weight of the two lures?

6. True or False: $\pi < 6$

7. True or False: $\pi^2 < 6$

8. Change $8\frac{2}{7}$ into an improper fraction.

9. $2^6 = ?$

10. Forty percent of Joe's clothes have fish stains on them. What percent don't have fish stains?

# The Bridge
from Chapters 1 to Chapter 7

fourth try

1. $3^4 = ?$

2. Darlene subscribes to Prof. Eldwood's *Magazine for Future Brides.* Since Joe likes to eat, she showed him the picture of the "Cone Cake" that was featured in the January issue. If you smooth out the frosting it would be a perfect cone with a height of 4 feet and a radius of 1½ feet. What is its volume? (Use $\pi = 3$.)

3. By showing him the bridal magazine, Darlene was hoping that Joe would "get the hint" and propose to her. Instead, he said, "That cake looks like the sinker you gave to me."
   "Speaking of sinkers," he continued, "tomorrow is the first of the month. That's always a good day to go fishing." Is *first* an ordinal number or a cardinal number?

4. "May I go with you?" Darlene asked. She really hated fishing, but she wanted to be near Joe. "I'll pack a nice lunch," she added. When she said that she'd bring the lunch, Joe said yes 100% of the time. What percent of the time did Joe say no?

5. Joe's favorite sandwich had $\frac{2}{5}$ lbs. of baloney and $\frac{2}{5}$ lbs. of ketchup. Together, how much did those two ingredients weigh?

6. The whole sandwich (with the baloney, the ketchup, the bread, the pickle relish, and the mayonnaise) weighed $3\frac{1}{4}$ lbs. Darlene cut it in half. How much did each half weigh?

7. True or False: $2\pi > 10$

8. Darlene read $1\frac{1}{8}$ bridal magazines per hour when she was on the boat with Joe. How many magazines did she read during the 6 hours she was on the boat? (Simplify your answer, if possible.)

9. Darlene asked Joe, "If we were to get married, where would you like to spend our honeymoon?" He wrote down the set of places he would like: {fishing at Lake Alpha, fishing at Lake Beta}. She wrote {fishing at Lake Alpha, fishing at Lake Beta, relaxing in Hawaii}. (She had copied his list and added Hawaii.) Are their two sets disjoint?

10. $\frac{3}{4} \times \frac{3}{4} = ?$

The Bridge
from Chapters 1 to Chapter 7

fifth try

1. On the fishing trip, Darlene brought along her collection of bridal magazines to read while Joe fished. She put them into a box that was 14 inches long, 12 inches wide and 8 inches deep. What was the volume of that box?

2. During the 50 minutes Joe and Darlene were in Fred's math class each day, Joe would think about fishing 12% of the time. How many minutes did Joe think about fishing?

3. In the math class, Darlene would sometimes think about the bride's maids she wanted at her wedding. She wrote: {Betty, Jennifer, Alicia, Misty, Georgette, Frankie}. What is the cardinal number associated with this set?

4. During the 50 minutes Joe and Darlene were in Fred's math class each day, one-fifth of the time Darlene would be thinking about what kind of bridal veil she wanted at her wedding. How many minutes was that?

5. $4^3 = ?$

6.

Medieval Bridal Hat with fake ears

In the May issue of Prof. Eldwood's *Magazine for Future Brides* Darlene read the article "Making Your Wedding Medieval." The cone-shaped hat had a radius of 4 inches and a height of 12 inches. What was its volume? (Use 3 for π.)

7. $\frac{4}{5} + \frac{3}{5} = ?$ Simplify your answer as much as possible. Do not leave it as an improper fraction.

8. $\frac{4}{5} - \frac{3}{5} = ?$

9. $\frac{4}{5} \times \frac{3}{5} = ?$

10. $\frac{4}{5} \div \frac{3}{5} = ?$ Simplify your answer as much as possible. Do not leave it as an improper fraction.

Chapter Eight
Tooth Brushing

Seeds! That was the next step. Fred knew that you couldn't just stick gobs of toothpaste in the ground and grow teeth. Toothpaste! Fred thought to himself. I haven't brushed my teeth this morning.

He raced down the hallway and found the door marked:

Years ago, when Fred first came to teach at KITTENS University, the janitor had added "*and one boy's*" to the sign for Fred's sake.

That sign still bothered Fred. He wanted to change "MENS" into "MEN'S," but he was too short to reach up and add an apostrophe. When you are 5½ years old and 36 inches tall, there are some things you can't do yet.

There was a short sink in the rest room that the janitor had installed just for Fred. On it lay his toothbrush, toothpaste, and dental floss. He flossed first and then brushed. Sometimes he used a little too much toothpaste.

Years ago when Fred got his first baby teeth, he knew he was supposed to brush them, but no one had ever shown him how to do it. He used a half a tube of toothpaste the first time he tried. He wanted his teeth to be "really clean."

The second time he only used a third of a tube of toothpaste. This could get a little expensive Fred thought to himself. But my teeth are worth it. I'm going to buy a case of toothpaste next time I'm at the store.

He had used a half a tube and a third of a tube. $\frac{1}{2} + \frac{1}{3}$

Chapter Eight Tooth Brushing

To add fractions, the denominators (the bottom numbers) have to be equal. We need a **common denominator**. Take the $\frac{1}{2}$ and multiply the top and bottom by 3. Then $\frac{1}{2}$ will turn into $\frac{3}{6}$.

Take the $\frac{1}{3}$ and multiply the top and bottom by 2. Then $\frac{1}{3}$ will turn into $\frac{2}{6}$.

Instead of $\frac{1}{2} + \frac{1}{3}$, we now have $\frac{3}{6} + \frac{2}{6}$. We have made the denominators alike.

$\frac{3}{6} + \frac{2}{6} = \frac{5}{6}$ Fred had used five-sixths of a tube of toothpaste.

He looked at the tube. It only had one-sixth left in it. He threw it away.

$$1 - \frac{5}{6}$$
$$= \frac{1}{1} - \frac{5}{6}$$
$$= \frac{6}{6} - \frac{5}{6}$$
$$= \frac{1}{6}$$

Today, Fred is a mature five-and-a-half year old. He only uses a pile of toothpaste about an inch tall on his brush.

Wait! I, your reader, have a question.

Yes. What is it?

Fred is doing all the flossing and brushing. But he hasn't had any breakfast yet. What's going on?

Fred sometimes forgets to eat. He never really "got into the habit," as he expresses it. That's one major reason he is five and a half and weighs only 37 pounds. But we know he likes to brush. Every month or two he heads to the store to buy a case of toothpaste.

But what's there to clean if he hasn't had breakfast?

Chapter Eight   Tooth Brushing

Why We Brush*

(Story board layout for the proposed film)

---

scene 1

Start with duck talking about the importance of good teeth.

---

scene 2

You don't brush just to get the pickle relish out of your mouth.

---

scene 3

The big enemy is plaque. (Rhymes with quack.)

Show war films of toothbrushes killing plaque.

---

scene 4

Plaque = thin film that sticks to teeth. Sugar & bacteria stick to plaque.

---

scene 5

THE BIG SCENE

The bacteria picnic on the plaque blanket eating the sugar.

---

scene 6

𝕳𝖔𝖗𝖗𝖔𝖗!

Bacterial excreta** is acid. Eats holes in teeth (called cavities).

✶ "Why We Fight" was a series of seven black-and-white films made during World War II. They were made to show the reason why the United States was involved in the war. **Why We Brush** is also in black-and-white.

✶✶ germ poop

55

Chapter Eight   Tooth Brushing

Your Turn to Play

1. If you were going to add $\frac{1}{6}$ and $\frac{3}{8}$ you could use 48 as the common denominator since both 6 and 8 divide evenly into 48. But 48 is not the *least* common denominator. What is the smallest number that both 6 and 8 evenly divide into?

2. Add $\frac{1}{6}$ and $\frac{3}{8}$

3. What is the least common denominator for $\frac{3}{10}$ and $\frac{11}{100}$?

4. Add $\frac{3}{10}$ and $\frac{11}{100}$

5. Check a box.
   When we have two mixed numbers such as $5\frac{1}{4}$ and $3\frac{1}{3}$ when do we need to change them into improper fractions?
   ☐ When we are going to add or subtract them.
   ☐ When we are going to multiply or divide them.

. . . . COMPLETE SOLUTIONS . . . .

1. The smallest number that 6 and 8 both evenly divide into is 24.

2. $\frac{1}{6} + \frac{3}{8} = \frac{4}{24} + \frac{9}{24} = \frac{13}{24}$

3. The smallest number that both 10 and 100 evenly divide into is 100.

4. $\frac{3}{10} + \frac{11}{100} = \frac{30}{100} + \frac{11}{100} = \frac{41}{100}$

5. The answer is *not* when we add them. To add $5\frac{1}{4}$ and $3\frac{1}{3}$ you add the whole numbers separately from adding the fractions.

$5\frac{1}{4} + 3\frac{1}{3} = 5\frac{3}{12} + 3\frac{4}{12} = 8\frac{7}{12}$

It's when you ☒ multiply or divide mixed numbers that you need to first change them into improper fractions.

$5\frac{1}{4} \times 3\frac{1}{3} = \frac{21}{4} \times \frac{10}{3} = \frac{210}{12} = 12\overline{)210} = 17\frac{6}{12} = 17\frac{1}{2}$

$\phantom{5\frac{1}{4} \times 3\frac{1}{3} = \frac{21}{4} \times \frac{10}{3} = \frac{210}{12} = 12\overline{)210}}\ \underline{-12}$
$\phantom{5\frac{1}{4} \times 3\frac{1}{3} = \frac{21}{4} \times \frac{10}{3} = \frac{210}{12} = 12\overline{)210}\ }\ \ 90$
$\phantom{5\frac{1}{4} \times 3\frac{1}{3} = \frac{21}{4} \times \frac{10}{3} = \frac{210}{12} = 12\overline{)210}}\ \underline{-84}$
$\phantom{5\frac{1}{4} \times 3\frac{1}{3} = \frac{21}{4} \times \frac{10}{3} = \frac{210}{12} = 12\overline{)210}\ \ \ }\ 6$

Chapter Nine
Gardening Mall Announced

Fred rinsed his mouth (5 minutes). Rinsed his toothbrush (3 minutes). Cleaned the toothpaste off his face (2 minutes). (It's a lot quicker if you don't use sooooooooooo much toothpaste.)

Fred was eager* to get back to his office to continue working on his victory garden. He ran down the hall and entered his office. He stood there and stared at the big cone-shaped pile of dirt. *I know I need seeds,* he thought to himself. *But I also need gardening things like shovels.*

He headed back out into the hallway, skipped down two flights of stairs, and walked outside. *But which way do I go?*

Just then, a newspaper boy came by with the latest edition.

Read all about it!
New gardening spot!

Fred gave him a dime.

The right way to write a dime 10¢ 10.¢ $0.10 ten pennies	The wrong way .10¢ (that's a tenth of a cent)

---

* English lesson: Before the middle of the 1700s, when you said that someone was *anxious* to do something, that meant that there was *anxiety* involved. *Anxiety* = worry or fear of danger. But lately (for the last couple of centuries) things have gotten a bit sloppy. People will say they are anxious to get started on a trip to Yosemite when they have no worry or fears attached to the trip.

Being an old fuddy-duddy, I like to say *I am anxious to go to the dentist, but I am eager to go and have pizza for lunch.*

Chapter Nine   Gardening Mall Announced

The KITTEN Caboodle

The Official Campus Newspaper of KITTENS University   Friday 6:40 a.m. Edition 10¢

Gardening Mall Grand Opening

New Mall

KANSAS: When the KITTENS University president announced earlier today, "We must plant, plant, plant!" construction began on the new mall. This mall is dedicated to every aspect of gardening.

The 80 stores were completed at 6 a.m. and were all rented out by 6:30 a.m. The Grand Opening will be at 6:45 a.m.

The new mall is located eight blocks north of KITTENS University.

Name-the-New-Mall Contest

This mall has been constructed so quickly that no one has yet thought of a good name for it. Everybody has just been calling it the New Mall.

The owner of the mall, C.C. Coalback, has announced a $75,000 first prize to the best name submitted by 7 a.m. today.

advertisement

Glenda's Garden Goodies
New Mall Special!

Backhoe
$1349

I've got five minutes to get there for the grand opening. Fred started jogging. *If I go two blocks per minute for four minutes, that will get me there.*

$$8 \text{ blocks} = \frac{2 \text{ blocks}}{\text{minute}} \times \frac{4 \text{ minutes}}{1}$$

distance equals rate times time

or as they like to write it in algebra

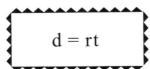

$$d = rt$$

Chapter Nine   Gardening Mall Announced

Some notes about d = rt

♪#1: d = rt is one of the most commonly used formulas in algebra.

♪#2: Joe once told Darlene that he liked to call d = rt the "dirt" formula. She asked him why. He said that d = rt can be spelled "d *iquels* rt."

    Darlene exclaimed, "But *equals* is spelled e-q-u-a-l-s."

    Joe answered, "But *dert* isn't a word."

♪#3: When you are multiplying fractions, you can cancel numbers. For example, at the end of Chapter 5 we saw

$$8 \times 7\frac{1}{2} = \frac{8}{1} \times \frac{15}{2} = \frac{\cancel{8}^{4}}{1} \times \frac{15}{\cancel{2}_{1}} = \frac{60}{1} = 60$$

Using **dimensional analysis**, we can also cancel dimensions:

$$8 \text{ blocks} = \frac{2 \text{ blocks}}{\text{minute}} \times \frac{4 \text{ minutes}}{1}$$

becomes

$$8 \text{ blocks} = \frac{2 \text{ blocks}}{\cancel{\text{minute}}} \times \frac{4 \cancel{\text{ minutes}}}{1}$$

$$8 \text{ blocks} = 2 \text{ blocks} \times 4$$

which everybody knows is true.

Dimensional analysis is sometimes called **unit analysis**.

    Fred got to the New Mall in four minutes. It was 6:44 a.m. Thousands of people were waiting for the grand opening. Fred couldn't believe it.

    He said to the man standing next to him, "I didn't know there were that many people interested in gardening."

    Gardening? I'm here to win the $75,000.

Your Turn to Play

1. Which of these is/are true?
                      .6¢ < a nickel
                      $0.10 < a dime
                      .98¢ < a quarter

2. If there were a thousand people who divided the $75,000 prize equally among them, how much would each one receive?

3. If you worked five hours and were paid $15 per hour, how much would you receive?

4. If you plant artichokes, you might get 8 artichokes for every square yard that you plant. How many artichokes would you expect if you planted 7 square yards?

5. Fred used a third of a tube of toothpaste to brush his teeth. The tube was half full when he started. How much was left in the tube?

6. Fred has 20 teeth, 8 of which are incisors. What fraction of his teeth are incisors? (Incisors are the flat cutting teeth in the front of the mouth.)

. . . . COMPLETE SOLUTIONS . . . .

1. .6¢ is six-tenths of one cent. It is less than a nickel.
$0.10 is the same thing as a dime, so it is not true that $0.10 < a dime.
.98¢ is $\frac{98}{100}$ of one cent. .98¢ is a little bit less than one cent, so it is true that .98¢ < a quarter.

2. $75,000 ÷ 1000 = $75,000. ÷ 1000 = $75. (To divide by a thousand, you move the decimal point three places to the left.)

3. 5 hours × $\frac{\$15}{\text{hour}}$ = $\frac{5 \text{ hours}}{1}$ × $\frac{\$15}{\text{hour}}$ = 5 × $15 = $75.

4. $\frac{8 \text{ artichokes}}{1 \text{ square yard}}$ × 7 square yards =

    $\frac{8 \text{ artichokes}}{1 \text{ square yard}}$ × 7 square yards = 8 artichokes × 7 = 56 artichokes

5. One-half take away one-third = $\frac{1}{2} - \frac{1}{3} = \frac{3}{6} - \frac{2}{6} = \frac{1}{6}$ of a tube.

6. $\frac{8}{20}$ which reduces to $\frac{2}{5}$

Chapter Ten
At the New Mall

Fred looked at the crowd of people who were waiting for the new mall to open. Many of them were holding pens in one hand and long lists in the other. *I feel silly. I don't have a list of the gardening supplies I need* Fred thought to himself.

It was 6:45 a.m.—time for the grand opening. The man in funny clothes was in charge of the opening ceremonies. "Welcome one and all to our new mall. It is dedicated to all your gardening needs. We hope that you will buy, buy, buy, so that you can plant, plant, plant."

"Hey!" someone in the crowd shouted. "The name-the-new-mall contest ends in 15 minutes. Where do we submit our entries?" He was waving a list he held in his hand.

The man in the funny clothes continued his speech, "This is such a grand occasion. I hope you will allow me a few moments to reflect upon the glorious history of this new mall and its prospects for a long and enduring future."*

"Cut the yacking!" another in the crowd yelled. "The contest ends at 7 a.m., and I am here to win. Where do I submit my entries?"

The owner of the mall, C.C. Coalback, had hired a circus ringmaster to lead the opening ceremonies. After a ten-minute speech, the ringmaster was supposed to introduce a juggling act.

This was Plan A. They were going to kill time until 7 a.m., and then no one could enter the contest. Coalback wouldn't have to pay out the $75,000 first prize.

It was time to switch to Plan B.

★ *long and enduring* = The guy in the funny clothes is padding his speech with unnecessary words. Some students, when they first are learning to write three-page papers, try to pad their papers with extra words. For example: `In studying the early Greek manuscripts and documents, it is important and necessary to think and reflect upon the words and statements made and written by the illustrious and famous Greeks from the long and distant past.`

Without the padding: `Early Greek writings deserve our attention.`

61

Chapter Ten   At the New Mall

The man in the funny clothes announced, "I see you are all in a hurry to shop at our wonderful and exciting new mall. Without further ado, I take great pleasure in being selected to be the one who may announce to the world that on this glorious day I, along with all my associates, now declare the opening ceremonies are now complete and done." (Which all could have been said with the words, "The mall is now open.")

The crowd surged past the ringmaster. But they were like many mobs: many arms, many legs, but no brains.

ringmaster

Sally, 2, came and stood on top of the fallen ringmaster. "Mister man, my mommy is going to win the contest."

Her mother came and pulled her away, saying, "How many times have I told you not to talk to people who are hurt!"

Sally said to her mom, "I was just gonna ask him where you can give in your list of names."

"Because of you, we've now lost your brother Jimmy." She yanked on Sally's arm, and they headed off.

The ringmaster was now all alone. Fred walked over to him and helped him stand up. There were footprints all over his fancy clothes. He was a little dazed and said to Fred, "I guess you want to know where to submit your name-the-mall entry."

"No sir," replied Fred. "I came to buy gardening supplies for my victory garden." Fred handed him a piece from an old KITTEN Caboodle newspaper to show he was serious about gardening.

Coalback's Plan B was simple. If the crowds wouldn't let the opening ceremonies go on for 15 minutes, he was going to have the ringmaster let them into the mall. Only one of the 80 stores would have the contest entry box . . . and that store wouldn't open until 7:01 a.m. Then Coalback wouldn't have to pay the $75,000 prize.

His Plan C was even more nefarious.* If someone had managed to somehow submit an entry, he would simply award the prize to himself.

✶ knee-FAIR-ee-us   Nefarious = extremely evil.

Chapter Ten   At the New Mall

"You're a good kid," the ringmaster told Fred. "Enjoy your visit to our mall. You had better do your gardening shopping today. I bet that this mall won't last very long." The ringmaster headed off to get his clothes cleaned.

Fred looked down the street of this outdoor mall. With eighty stores to choose from, he didn't know where to start shopping for his victory garden supplies. He looked at the mall directory map:

> Abby's Agricultural Supplies
> Beatrice's Better Blooms
> Cameron's Canning Supplies
> Dakota's Digging Equipment
> Eve's Garden
> Ferdinand's Farming
> Glenda's Garden Goodies
>   . . .

It was 7:01 a.m. and the mall was getting strangely quiet. Almost all of the thousands of people had left. There were pieces of paper littering the street. Fred picked one of them up. It was a list of possible names for the new mall.

"Look Mom!" said Jimmy, as he and his sister were being pulled along by their mother. "There's the scarecrow I told you about."

"That's not a scarecrow," his mother corrected him. "That's just a deformed kid with a square head. How many times have I told you not to talk about people who look weird?"

Sally, 2, who was in her mother's other hand, turned and winked at Fred as they passed by.

Fred passed the first storefront.

Abby's Agricultural Supplies

Sale!
20% off our best shovel.
Normally $18.

Chapter Ten   *At the New Mall*

Your Turn to Play

1. If Abby's shovels are normally $18, and you get a 20% discount, how much is their sale price?

Looking Back

If you take 20% off of a price,
that leaves 80% that you have to pay.

2. With a new Abby shovel, you can dig 19 cubic feet per hour. How many cubic feet could you dig in 3 hours?

3. If you have a new Abby shovel and can dig 19 cubic feet in the first hour, why couldn't you dig 190 cubic feet in the next ten hours?

4. At Beatrice's Better Blooms they had a sack of Better Bloom fertilizer. It was a 40-pound sack, but it had had a leak and lost 17% of its contents. How much did the sack now weigh?

5. At Cameron's Canning Supplies, they offered a jar that was two-thirds full of peach jam. It was a jar that could hold three-fourths of a quart. How much peach jam was in the jar?

·······**COMPLETE SOLUTIONS**·······

1. 80% of $18 = 0.80 × $18 = $14.40

2. $$\frac{19 \text{ cubic feet}}{1 \text{ hour}} \times \frac{3 \text{ hours}}{1}$$

$$= \frac{19 \text{ cubic feet}}{1 \, \cancel{\text{hour}}} \times \frac{3 \, \cancel{\text{hours}}}{1}$$

$= 19 \text{ cubic feet} \times 3$

$= 57 \text{ cubic feet}$

3. Because you would get tired.

4. If it lost 17% of its weight, it retained 83% of its weight.
83% of 40 pounds = 0.83 × 40 = 33.2 pounds.

5. Two-thirds of three-fourths = $\frac{2}{3} \times \frac{3}{4} = \frac{\cancel{2}^1}{\cancel{3}_1} \times \frac{\cancel{3}^1}{\cancel{4}_2} = \frac{1}{2}$ quart of jam

Chapter Eleven
Seed Money

Coalback had never read the Kansas laws. "Why read them," he used to laugh and say, "if you are not going to obey them?" That almost made sense.* What he ignored was the fact that he had to obey those laws regardless of whether or not he had read them.

There may be a law in Kansas that states:

§1302 Kansas Fair Contest Law

¶ 293.1 Any public offering of any prize over $100 shall be deemed a contest under this section.

¶ 293.2 All contest prizes must be awarded if there are any entries to that contest.

¶ 293.3 Any entry submitted to any owner, employee, agent, or attorney of the holder of a contest, shall be considered a valid entry.

¶ 293.4 No prizes shall be awarded to any owner, employee, agent, or attorney of the holder of a contest.

When Coalback was questioned by the Grand Jury, he admitted that he had held a contest (¶ 293.1).

He testified that he had not awarded the $75,000 first prize because there were no entries. He said that was totally legal under ¶ 293.2.

∗ A lot of evil things almost make sense. That's what helps them become popular. If you go around saying 2 + 2 = 32906235, nobody is going to believe you. But if you say things that are almost true, you can get people to accept the lies hidden in your statements.

Example: Vacuum cleaner salesman says: ① You want clean rugs. ② I have a machine that will give you the cleanest rugs possible. ③ Do you deserve the cleanest rugs possible? ④ You can afford $40 per month. What he fails to mention is that the vacuum cleaner costs $6,982.

Example: The Leader proclaims: ① The people of our country are a great people. ② They should not be denied anything that they need. ③ Our people need living space. ④ We can take living space from the country bordering our country. In 1939, the people believed those four almost-true statements of the Leader and his evil conclusion: We should cross the border and take their land.

Chapter Eleven   Seed Money

When the ringmaster was called to testify, they asked him if he had received any entries in the contest. He said, "Just this scrap of paper that some kid had given me. This was entered into evidence.

When C.C. Coalback was called to testify again, he said, "It had slipped my mind. I now clearly remember awarding the $75,000. I was the one who won the contest."

The Grand Jury informed him that that was illegal by ¶ 293.4. They directed him to pay the $75,000 first prize to the only valid entry in the contest.

I'm sure I'll need a shovel, Fred thought to himself. He walked into Abby's Agricultural Supplies. The place was a mess.

Wow. This place could sure use Carolus Linnaeus, Fred thought. He was too polite to mention that to Abby when she came up and asked, "Is there anything I can help you find?"

Chapter Eleven Seed Money

Fred answered, "Yes. I'd like a shovel and some seeds."

Abby found the shovel. It was right next to the cactus plants. But after ten minutes of looking, she admitted, "I thought the seeds were right behind the chickens. But I can't seem to find them. I wish that Carolus Linnaeus were here."

"That's funny," said Fred. "I was thinking about Linnaeus also." [pronounced leh-KNEE-us]

There was a large picture of him on the wall behind Abby.

Abby laughed and said, "I guess it's always good to put up a picture of someone you wish you could imitate."

Carolus Linnaeus

They both understood the reason for her laughter.

Intermission

For the rest of you who don't get the joke . . .

Carolus was Mr. Organization when it comes to biology. He loved to classify every living thing. When he was 25, he walked all over Lapland collecting plants to classify. He covered 4,600 miles, which he did on foot.

Then on to Germany, Holland, England and France. At no point did he ever ride in a car, truck or bus. (He was born in 1707.)

His classification scheme for every plant and animal is used everywhere in the world today.

Every species has been assigned two names. Once you say the two magic words, you have uniquely identified that species.

For example, if you say *Zea mays*, everyone in Philadelphia or in the Phillippines knows that you're talking about that yellow stuff with kernels (and maybe a dab of butter).

used everywhere

If this is your book that you bought with your own money and you have a yellow crayon, you know what to do.

Chapter Eleven   Seed Money

Your Turn to Play

1. Fred dreamed of planting some *Zea mays* in his victory garden. If he planted it in soil that was 2 feet deep, and if the height of his office is 8¼ feet tall, how high could his corn grow before it hit the ceiling?

2. Fred dreamed that if someday he were rich, he would order the complete collection of Professor Eldwood's books. They would come in cartons which are 1¾ feet tall and would be stacked outside of Fred's office in the hallway (until Fred got the dirt out of his office). The hallway is also 8¼ feet tall. How many cartons could be in a stack?

.......**COMPLETE SOLUTIONS**.......

1. The **General Rule** is: *If you don't know whether to add, subtract, multiply or divide, first restate the problem with really simple numbers.* If, for example, you had 2 feet of soil and the room was 8 feet tall, you know the plant could be 6 feet tall. *Look to see how you got the 6 feet*—you subtracted—*and then you know what to do in the original problem.*

$$8¼ - 2 = \quad \begin{array}{r} 8\ ¼ \\ -\ 2\ \\ \hline 6\ ¼ \end{array}$$

The plant can grow 6 ¼ feet tall before it hits the ceiling.

2. Using the **General Rule**: *If you don't know whether to add, subtract, multiply or divide, first restate the problem with really simple numbers,* we suppose the cartons were 2 feet tall and the hallway was 8 feet tall. Then you could stack 4 cartons. How did you get the 4? You divided.

So you use division in the original problem.

$$8¼ \div 1¾ = \frac{33}{4} \div \frac{7}{4} = \frac{33}{4} \times \frac{4}{7} = \frac{33}{\cancel{4}} \times \frac{\cancel{4}^1}{7} = \frac{33}{7} = 4\frac{5}{7} \text{ cartons.}$$

If you tell someone that you can stack $4\frac{5}{7}$ cartons, they may think you are a little weird. The most that you can stack is 4 cartons.

Chapter Twelve
Getting Things Arranged

Fred offered to help Abby straighten up her store. She said that she wasn't sure where to begin. There were so many ways they could arrange all the items in the store.

They could arrange everything alphabetically. *At one end of the store start with the apple peelers, the bamboo rakes, the cabbages.*

They could arrange everything by size. *At one end of the store start with the seeds, and at the other end would be the backhoes.*

They could sort everything by their colors. *At one end of the store put the red items, then the blue items, etc.*

They could sort everything by price. *At one end of the store start with the one-cent items.*

They could arrange everything by function. *At one end of the store start with the things that dig, such as shovels. Then put the things that scratch the soil, such as rakes and chickens.*

"Carolus Linnaeus had a tougher job than what we've got," Fred said. "He had the whole world of living things that he had to sort out."

"How did he ever start?" Abby asked.

"First, he divided everything into two kingdoms: the plant kingdom and the animal kingdom—Kingdom Plantae and Kingdom Animalia, to use the official names." Fred was simplifying a little bit. Everyone knows that there are five kingdoms.

Wait! Stop! I, your reader, haven't the faintest idea what you are talking about. Everything living has got to be either a Plantae or an Animalia—a plant or an animal.

No.

What do you mean? You have got to be fooling. Are you going to invent a Kingdom Cartoonia for Donald Duck?

No. I'm serious. There are five kingdoms for living things.

★ Kingdom Monera—single-celled bacteria. They have little internal complexity.
★ Kingdom Protista—single-celled or multi-celled with great internal complexity as compared to bacteria.
★ Kingdom Fungi—multi-celled that feed by extracellular digestion.
★ Kingdom Plantae—plants.
★ Kingdom Animalia—animals.

"Then he split each kingdom into phyla. Corn is in the kingdom Plantae and the phylum Anthophyta. Anthophyta means flowering plants," Fred continued.

Hold it! This sounds like Latin. Is phyla the plural of phylum?

There is a reason it sounds like Latin. It is. Even the Swede named Carl von Linné who was born in 1707 changed his own name into the Latin form, Carolus Linnaeus.

Fred was on a roll and couldn't be stopped: "Then Carl split each phylum into classes
   and each class into orders,
      and each order into families,
         and each family into genera,
            and each genus into species."

The rules for writing these classifications:
   (1) Capitalize kingdoms, phyla, classes, orders, families, genera, but not species.
   (2) Italicize genera and species.

Corn is in the genus *Zea* and in the species *mays*. No other plant (or animal) is called *Zea mays*.

So, I, your reader, now have to say at the dinner table, "Please pass the Zea mays"? My brothers and sisters are going to laugh at me.

But if you are in England, "maize" is the word they use for "corn." You are supposed to say, "Please pass the maize."

And what if I'm in England and I say, "Please pass the corn"? Won't they understand what I'm talking about?

Chapter Twelve   Getting Things Arranged

You'll get a real surprise. If you say, "Please pass the corn," they will hand you wheat! And in Scotland, they will hand you oats!
Why don't they learn to speak English?
They ask the same thing about us.

So Fred and Abby put all the tools in one section of the store, all the seeds in another section, and all the chickens in a third section.

Your Turn to Play

1. The *Zea mays* seeds came in 50-pound sacks. Fred, who weighs 37 pounds, had a little trouble lifting them. He looked around and saw Abby's forklift. No problem, he thought. He climbed in and started up the motor. At the age of five, no one had ever told him that forklifts are designed to pick up stuff on wooden pallets.
   He zoomed across the store and speared one of the 50-pound sacks.

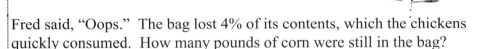

Fred said, "Oops." The bag lost 4% of its contents, which the chickens quickly consumed. How many pounds of corn were still in the bag?

2. Fred pulled 2¼ feet of tape off of Abby's tape dispenser and put it on the leaking seed bag. Then he pulled off another strip which was 1⅔ feet long and made a big ✕ on the bag with the tape. How much tape did he use?

.......**COMPLETE SOLUTIONS**.......

1. If it lost 4% of its contents, 96% were still in the bag.
96% of 50 pounds = 0.96 × 50 = 48 pounds were still in the bag.

2. $2\frac{1}{4} + 1\frac{2}{3} = $ $\begin{array}{r} 2\frac{1}{4} \\ + 1\frac{2}{3} \\ \hline \end{array}$ $\begin{array}{r} 2\frac{3}{12} \\ 1\frac{8}{12} \\ \hline 3\frac{11}{12} \end{array}$ (or 3 feet, 11 inches)

Chapter Thirteen
Blood and Your Brain

Fred and Abby had finished straightening out her store. The hoes and rakes were with the shovels. The roses were with the carnations. The chickens were with the. . . .

"Where'd the chickens go?" asked Abby.

"I put them over with the ducks," said Fred.

"They're not there," said Abby.

They looked all over the store. The chickens were gone. Then they looked outside in front. The chickens were in the street eating the lists of names for the new mall that people had dropped.

The street was empty except for a car with blue and red flashing lights.

"I think we're in trouble," Abby whispered to Fred.

A very large police officer got out of the car and approached them.

"I'm sorry, officer," Abby began. "The chickens just escaped when we left the front door open." Meanwhile, Fred was running around gathering up chickens and stuffing them back into the store.

"Hey kid," he said, pointing to Fred.

"Who me?" Fred turned white as a sheet of paper.

The policeman looked carefully at Fred and asked, "Are you the one who gave this scrap of paper to the ringmaster?" He held it out for Fred to look at. Fred swallowed hard and nodded. Speech had left him. I didn't know that you weren't supposed to hand ringmasters a piece of paper Fred thought to himself. I hope he is not going to arrest me.

Chapter Thirteen   Blood and Your Brain

"You seem to fit the description and the picture," the policeman said as he handed Fred a check.

```
C.C. COALBACK, OWNER                              1001
NEW MALL
                         Date  Today
Pay to the
order of   The Contest Winner        $ 75,000.00
           Seventy-five thousand and 00/100      bucks

           Kittens Bank

                                  C.C. Coalback
"IT'S IN THE KITTY"
```

Fred did the appropriate thing. He passed out. Luckily, the chickens were all back in Abby's store, so they didn't come and eat Fred's check.

▷▷▷ From Prof. Eldwood's *Guide to Fainting*, 1837 ◁◁◁

There are certain places you don't want to have blood. E.g., you don't want it all over the living room rug.* But there are other places where blood is really good. E.g., your brain likes a lot of nice, fresh, red blood. When blood passes through the lungs, it picks up oxygen and turns bright red. Then one trip through the heart and then up to the brain.

If something messes up this trip, your brain doesn't like it. You faint.

E.g., if you spend a couple of days in bed and then suddenly get up, you might have a temporary drop in blood pressure and feel faint.

E.g., an emotional shock (such as receiving a check for $75,000) can affect the nerves that control blood pressure and cause you to feel faint.

There are 298,985 other possible reasons your brain might not be receiving all the nice, fresh, red blood it likes.

---

* Two things to note: In 1837, they typically didn't have wall-to-wall carpets. They had rugs. Second, "E.g." means "for example."

Chapter Thirteen   Blood and Your Brain

Prof. Eldwood's First-Aid Quiz

Suppose someone is feeling faint or has fainted. Which is the better position for them to be in?

Alternative A: Standing up.

— Note blood in lower half of body.

Alternative B: Lying down with legs raised.

ANSWER: It all depends on where the person's brain is located.

Your Turn to Play

1. Sometimes comparing two numbers is easy. E.g., \$75,000 > \$1. At other times, it can be a little more difficult. Which is larger: $\frac{2}{3}$ of a pound of chocolate or $\frac{7}{12}$ of a pound of chocolate? (Hint: to compare two fractions, the trick is to first make their denominators alike—*as if* you were going to add the fractions.)

2. If an ounce of chocolate contains $30\frac{1}{4}$ calories, how many calories are in $2\frac{1}{11}$ ounces?

. . . . COMPLETE SOLUTIONS . . . .

1. If we were going to add $\frac{2}{3}$ and $\frac{7}{12}$ the common denominator would be twelve. So asking if $\frac{2}{3}$ is greater than $\frac{7}{12}$ is the same as asking if $\frac{8}{12}$ is greater than $\frac{7}{12}$. That's easy. The answer is yes.

2. Remember the **General Rule**: *If you don't know whether to add, subtract, multiply or divide, first restate the problem with really simple numbers.* If an ounce contained 3 calories and you had 4 ounces, your brain tells you that you have 12 calories. *You multiplied.* Now we do it with the real numbers. $30\frac{1}{4} \times 2\frac{1}{11} = \frac{121}{4} \times \frac{23}{11} = \frac{\overset{11}{\cancel{121}}}{4} \times \frac{23}{\underset{1}{\cancel{11}}} = \frac{253}{4} = 63\frac{1}{4}$ calories.

Chapter Fourteen
Conversion Factors

Abby raised Fred's legs, and he was soon awake. Fred lay there for a moment. Sometimes, when your world changes radically, it takes a minute or two to get adjusted. Fred's salary at KITTENS University was $500 per month. The check in his hand was for $75,000. How many month's wages was that? he thought to himself.

Fred knew about the **General Rule*** since he taught that rule in his math classes. He restated the problem as *Suppose I were making $2 a month, and the check was for $10.* Without having to do any deep thought, everyone knows that $10 is five month's wages. *You divided to get the answer.* So to find out how many month's wages (at $500/month) the check for $75,000 represented, Fred divided.

$$\begin{array}{r} 150 \\ 500\overline{)75000} \\ -500 \\ \hline 2500 \\ -2500 \\ \hline 0 \end{array}$$

So that check represented 150 months of working at KITTENS. How many years is that?

One way to figure out how many years is 150 months is just divide by 12. Another way is to use a **conversion factor**.

$$150 \text{ months} \times \frac{1 \text{ year}}{12 \text{ months}} = 150 \cancel{\text{months}} \times \frac{1 \text{ year}}{12 \cancel{\text{months}}} = \frac{150 \text{ years}}{12}$$

$$= 12\overline{)150.0} \begin{array}{r} 12.5 \\ -12 \\ \hline 30 \\ -24 \\ \hline 60 \\ -60 \\ \hline 0 \end{array}$$

So 150 months equals 12½ years of wages. This Sunday, Fred's Sunday School offering would contain an extra $7,500.

---

* If you don't know whether to add, subtract, multiply or divide, first restate the problem with really simple numbers.

Chapter Fourteen   Conversion Factors

A conversion factor is a fraction whose numerator (top) is equal to its denominator (bottom). So a conversion factor fraction is always equal to one. And multiplying by one is always okay.

Suppose you had 43 acres you wanted to plant with watermelon seeds, and you wanted to know how long it would take you to do that planting.

Yesterday you found that you could plant 5.22 acres in 7 hours. This gives you your conversion factor since 5.22 acres = 7 hours. The conversion factor will either be $\frac{5.22 \text{ acres}}{7 \text{ hours}}$ or it will be $\frac{7 \text{ hours}}{5.22 \text{ acres}}$

Which one? You pick the conversion factor *so that the units cancel.*

$$\frac{43 \text{ acres}}{1} \times \frac{7 \text{ hours}}{5.22 \text{ acres}} = \frac{43 \times 7 \text{ hours}}{5.22} = \text{etc.}$$

"I almost forgot why I came into your store," Fred said. "I saw your sign. You have shovels on sale for $14.40. I'd like to buy one."

Abby smiled. "That shovel may be a little too big for you. Here's one that may be more your size." She handed him a plastic shovel that she had taken out of child's sand pail.   "It's only 69¢."

Fred handed her the $75,000 check and she gave him his change.

```
  75000.00
−     .69
  74999.31
```

Fred put the penny, nickel and quarter (= 31¢) into his pocket. He put the shovel and all the paper money into a paper bag.

Wait! Stop! I, your reader, can't stand it. This is crazy. Where did Abby get all that cash? Stores don't usually carry that much change.

I don't know either. I do know that little five-year-old boys usually don't carry around almost $75,000 in cash in a paper bag. But you just

76

Chapter Fourteen   Conversion Factors

saw him get that money and put it into the bag. The older you get, the more you learn that this world is not only stranger than you might have imagined, but possibly stranger *than you could imagine.**

Your Turn to Play

1. If you had $75,000 in hundred dollar bills, how many bills would that be? Use the **General Rule** approach: *If you don't know whether to add, subtract, multiply or divide, first restate the problem with really simple numbers.*
2. If you had $75,000 in hundred dollar bills, how many bills would that be? This time, use the conversion factor approach.
3. If you had $75,000, and you had to pay 40% of it for taxes, how much would you have left?

. . . . **COMPLETE SOLUTIONS** . . . .

1. Simplifying to easy numbers, I suppose I have $600 in hundred dollar bills. My brain tells me without thinking that I'd have six bills. *I divided $600 by 100.* So I should divide $75,000 by a hundred.
   Dividing by a hundred moves the decimal two places to the left. 75,000 → 75,000. → 750.   I would have 750 bills.
2. $100 = one bill
So my conversion factor will either be $\frac{\$100}{\text{one bill}}$ or $\frac{\text{one bill}}{\$100}$

I choose the one that makes the units cancel.

$$\frac{\$75,000}{1} \times \frac{\text{one bill}}{\$100} = \frac{\cancel{\$}75,000}{1} \times \frac{\text{one bill}}{\cancel{\$}100} = \frac{75,000 \text{ bills}}{100}$$

= 750 bills.
3. If you paid 40% in taxes, you would keep 60%. 60% of $75,000 = 0.6 × 75,000 = $45,000.

―――――――――――

* The physicist, Sir Arthur Eddington, said something very similar to this. Eddington also wrote, "Something unknown is doing we don't know what."
   At least in mathematics, things may be a little more straightforward. If you have to add two fractions, the first thing to do is make the denominators alike.

77

Chapter Fifteen
Plants Don't Eat Dirt

Fred headed outside. He looked at the mall directory again and located Glenda's Garden Goodies. He remembered that they were having a sale on celery seeds.

A mother was dragging her two little kids along. She smiled at Fred and said, "Good day sir."

Her little boy Jimmy said, "Mommy, I thought you told me not to talk to people that looked weird. You just said hello to him."

"He's not really weird. He has a bag full of money," she explained.

Two-year-old Sally turned and winked at Fred as they passed by.

Fred raced down the street until he came to

Fred walked in.

"Welcome," Glenda shouted. "You are our very first customer."

"That seems strange," Fred commented. "The mall has been open for almost an hour, and there were thousands of people here at the grand opening."

"They all came for the name-the-mall contest. Only you and a mother with two small kids are left in the mall. I'm worried that with no customers, we'll all have to close up our shops."

Fred tried to brighten her day and asked about the packet of celery seeds that she had advertised for 15¢.

"Here. Take it kid," she said as she handed him the packet of seeds. "Fifteen cents isn't really going to help pay the rent. The mall owner, C.C. Coalback, told us that there would be an uncountable number of customers, and we would have no trouble paying the rent of $400."

Chapter Fifteen   Plants Don't Eat Dirt

"$400 per month is not a very high rent," Fred said.

"No, it isn't. But the rent is $400 per day." There was a little tear in Glenda's eye. "I believed him. He looked so honest."

Fred put the packet of celery seeds into his paper bag.

Glenda continued, "A lot of gardeners buy plants that have already germinated. Then they don't have to wait for the seeds to sprout. Here is a lovely little plant that I started last year. I can sell it to you for $2."*

The plant was much taller than Fred. Glenda explained to Fred that she would keep the container and the dirt, and he would get just the plant.

"Even then," she said, "That's really a bargain for $2. The plant weighs 15 pounds. After the plant is sold, I put in another seed and grow a new plant. I've done that for years with this same pot and soil. The only thing I add is an ounce of fertilizer when I start the new plant."

Fred blinked.

His heart skipped a beat.

His mouth dropped open.

His brain went *cha-ching!*

Fred was experiencing an epiphany.**

15-lb. plant

Perhaps, **the heart of biology** was wrapped up in what Glenda had been telling Fred about her growing her 15-pound plants.

✶ When I wrote this originally, I typed, "I can sell it to you for $2.." I thought this made sense. One period for the decimal point, and one period for the end of the sentence. The people who know English said that that was wrong. It should be, "I can sell it to you for $2."

English is a lot harder than math.

In math, for example, when you divide fractions $\frac{a}{b} \div \frac{c}{d}$ it is real simple. You always invert the one on the right and multiply $\frac{a}{b} \times \frac{d}{c}$

✶✶ [ee-PIFF-an-ee] There are several meanings of *epiphany*. In this context, Fred was suddenly understanding something important. (In other contexts, *epiphany* can mean a sudden appearance of God.)

Chapter Fifteen   Plants Don't Eat Dirt

Here are the thoughts that hit Fred (almost as hard as a $75,000 check).

THOUGHT #1: That 15-pound plant of Glenda's is partly water and partly "solid stuff."

THOUGHT #2: If I put that plant into an oven and dried it out, it would probably weigh around two pounds.

THOUGHT #3: If that two pounds of "solid stuff" came from the dirt, then Glenda's pot of dirt would get lighter and lighter with each new plant she grew. But it isn't lighter.

THOUGHT #4: If you plant the most massive tree in the world* out in the middle of some level ground and watch it grow, it doesn't eat up the dirt around it and end up in a hole.

When you walk in the forest you see this...

...not this!

THOUGHT #5: "Something unknown is doing we don't know what," but they are not getting most of their solid food from the dirt.

Quiz for You, the Reader (ANSWER IS ON THE NEXT PAGE.)

Most of the non-water mass of plants comes from . . .
    A) No one knows.
    B) Dust from fairies.
    C) Thin air.
    D) Plants don't grow. They just appear to grow.

★ The most massive tree in the world is the giant sequoia with the official name *Sequoiadendron giganteum*. (Following the spelling rules from page 70: Italicize genera and species. Capitalize the genus, but not the species.)

The giant sequoia can reach a height of 325 feet with a 30-foot trunk diameter. That's a lot of wood.

Chapter Fifteen   Plants Don't Eat Dirt

Historical fact: Before the early 1600s, there were many centuries. During those many centuries, it was common knowledge that plants "ate" dirt and used the dirt-food to grow up to be big and strong plants.

In the early 1600s, Jan Baptist van Helmont did a science project. He took 200 pounds of dry dirt and planted a five-pound willow tree in the dirt. Five years later, he pulled the tree out of its container and scraped off the dirt. The tree weighed 164 pounds. The dirt weighed two ounces less.

Your Turn to Play

1. There are 16 ounces in a pound. How much did the van Helmont's dirt weigh after his experiment?

2. If I weigh 180 pounds, 3 ounces, and I lose 7 ounces, how much will I weigh?

3. $\left(\frac{6}{7}\right)^3 = ?$

4. $2^? = 8$

- - - - - - - **COMPLETE SOLUTIONS** - - - - - - -

1. 200 pounds 0 oz.   199 pounds 16 oz.
    −   2 oz.   −   2 oz.
    ─────────────────   ─────────────────
                               199 pounds 14 oz.

2. 180 pounds 3 oz.   179 pounds 19 oz.
    −   7 oz.   −   7 oz.
    ─────────────────   ─────────────────
                               179 pounds 12 oz.

3. $\left(\frac{6}{7}\right)^3 = \frac{6}{7} \times \frac{6}{7} \times \frac{6}{7} = \frac{216}{343}$

4. $2^3 = 8$ since $2 \times 2 \times 2 = 8$

The answer to the quiz from the previous page is *C) Thin air*.

We'll explain that answer after you cross **The Bridge**.

81

The Bridge
from Chapters 1 to Chapter 15

first try

> Goal: Get 9 or more right and you cross the bridge.

1. Darlene and Joe were out in Joe's boat. Joe was fishing while Darlene made Joe's favorite baloney-and-ketchup sandwich. When she gave it to him, he asked, "Where's the butter?" She handed him a cube of butter that was three-quarters of an inch by three-quarters of an inch by three inches. What was its volume?

2. Joe put the cube of butter into his sandwich and smashed it flat between his hands. Before he smashed it, the sandwich weighed 2 lbs., 4 ozs. When he smashed it, 6 ozs. of ketchup dribbled out onto his lap. How much did the sandwich weigh after he smashed it?

3. Darlene didn't like to watch Joe eat. She read her Prof. Eldwood's *Magazine for Future Brides* instead. She read at the rate of $1\frac{2}{3}$ pages per minute. How long would it take her to read the whole 500-page magazine?

4. What is the volume of this cone? (Use 3 for π.)

5. Which is larger: $\frac{7}{12}$ or $\frac{3}{4}$?

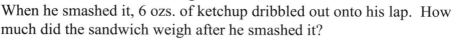

6. One of the articles in her *Future Brides* magazine was "How to lose 3.2 Pounds per Week Eating Radishes." How many pounds could she lose in five weeks?

7. Darlene thought to herself, "If this diet works, I could weigh zero pounds in 40 weeks." How much does she weigh now?

8. The *Future Brides* magazine had 500 pages in it. Twenty percent of those pages had pictures of bridal dresses. How many pages in the magazine didn't have pictures of bridal dresses?

9. Change $7\frac{3}{8}$ into an improper fraction.

10. $4^3 = ?$

The Bridge
from Chapters 1 to Chapter 15

second try

1. Every time Joe went fishing, he lost 8% of his fishing lures. Today he started with 400 lures. How many would he have at the end of the day?

2. Joe likes big fishing hooks. Really big. He felt it made him look more manly to own fishing hooks that weighed $3\frac{3}{4}$ pounds. He brought 7 of them along. Darlene asked him, "Are you planning on catching whales or something?" How much did those 7 hooks weigh?

3. The biggest fish that Joe had ever caught weighed $\frac{2}{3}$ of a pound. How much heavier was one of his big fishing hooks than that fish?

4. True or False? 0.10¢ < a nickel

5. Darlene weighs 128 pounds. Does she weigh more or less than 34 of Joe's $3\frac{3}{4}$ pound fishing hooks?

6. We are going to use the d = rt formula. If Darlene can row Joe's boat at the rate of 1.6 miles per hour and she rows for 0.7 hours, how far would they go?

7. If Joe loses $5\frac{1}{7}$ yards of fishing line per hour, how many would he lose in 10 minutes (which is $\frac{1}{6}$ of a hour)?

8. $\frac{3}{7} + \frac{4}{77} = ?$

9. Sixty-three percent of the time Joe and Darlene go fishing, Joe forgets to wear sun screen (and, of course, gets a sunburn). What percent of their fishing trips does Joe remember to use sun screen?

10. True or False? $1\frac{1}{4} \times 4 > 2\pi$

The Bridge
from Chapters 1 to Chapter 15

third try

---

1. Set A = {radishes, red, rink, roses}.
   Set B = {blue, red, green}.
   Set C = {boat, blue, balloon}.
   Set D = {radishes, green, balloon}. Is any pair of these sets disjoint? If so, which pair?

2. (Continuing the previous question) What is the union of B and C?

3. What is one-third of $8\frac{1}{3}$?

4. If a can of fishing worms cost $1.80, how much would seven cans cost?

5. The June issue of *Future Brides* magazine is really large. It measures $8\frac{1}{2}$" by 11" by 4". (" means *inches*) What is its volume?

6. $\left(\frac{7}{8}\right)^2 = ?$

7. Darlene bought a new tube of pink lipstick. She used $\frac{3}{16}$ of it on a fishing trip with Joe. How much was left?

8. True or False? 0.50¢ < ten pennies

9. If Joe uses an average of $7\frac{1}{4}$ worms for every hour he fishes, how many worms would he be expected to use on an 8-hour fishing trip?

10. If 60 *Zea mays* seeds weigh 2.3 grams, how much would 66 seeds weigh? (One way to do this problem is to use a conversion factor.)

The Bridge
from Chapters 1 to Chapter 15

fourth try

---

1. Joe used $4\frac{2}{3}$ pounds of worms in order to catch $1\frac{1}{4}$ pounds of fish. Darlene told him, "Wouldn't it be better if you just ate the worms?" How much more did the worms weigh than the fish?

2. There were two small holes in Joe's boat. He used $1\frac{1}{5}$ feet of tape to repair the first hole and $4\frac{1}{6}$ feet of tape to repair the second hole. How much tape did he use to make the two repairs?

3. Joe was wearing a new pair of fishing boots that he had just purchased. He had borrowed $20 from Darlene and spent 83% of that on those boots. How much did he spend?

4. The boots leaked "only a little." He tore off a $17\frac{1}{2}$ inch piece of tape and used a third of it to patch the boots. How much did he use?

5. How much of that $17\frac{1}{2}$ inches did he *not* use?

6. Which weighs more: $\frac{2}{9}$ of a pound of fishing lures or $\frac{1}{6}$ of a pound of pink lipstick?

7. Using 3.1 for π, what is the volume of a cone that is 10 miles high with a radius of 4 miles?

8. One yard equals 91.44 centimeters. Darlene rowed the boat 7 yards. How many centimeters was that? (One way to do this problem is to use a conversion factor.)

9. Change $6\frac{2}{9}$ into an improper fraction.

10. $\left(3\frac{2}{3}\right)^2 = ?$ Do not leave your answer as an improper fraction.

The Bridge
from Chapters 1 to Chapter 15

fifth try

1. $\left(\frac{2}{5}\right)^3 = ?$

2. Seventy-three percent of the value of everything that Joe owns is invested in his fishing gear. What percent is not invested in his fishing gear?

3. Joe dreamed of catching a whale some day. He figured that would make him famous since no one had ever caught a whale in Lake Alpha before. He told Darlene that if he caught a thousand-pound whale, he would give one-eighth of it to her. How much would she get?

4. Joe asked Darlene, "If I caught a big whale, would you cook a whale steak for me?" Darlene said she would, knowing that there are no whales in Lake Alpha.

    If she started with a 5 lb., 3 oz. piece of whale and she cut off the fat (which weighs 3 lbs., 7 oz.), how much would be left?

5. "And if I caught a whale," Joe said, "we could have a party and invite Fred, Betty, and Alexander to come." One serving of whale steak has 980 calories. How many calories are in five servings?

6. Which weighs more? A half-ton whale or Darlene's collection of bridal magazines that weighs $\frac{9}{16}$ of a ton?

7. $\frac{7}{8} - \frac{3}{4} = ?$

8. Joe dreamed of selling whale meat. It normally sells for $9/lb., but he imagined selling it for 40% less than that, because he was worried that it would all rot before it was sold. What price would Joe sell his whale meat for?

9. $\frac{7}{8} \div \frac{3}{4} = ?$ Do not leave your answer as an improper fraction.

10. $1^{3972} \times 0^{23955} = ?$

Chapter Sixteen
Out of Thin Air

Some things are easy to believe. After all, how did all those Christmas presents get under the tree if some 250-pound guy in a red suit didn't come down your 8-inch diameter chimney? It's only when you are able to do basic arithmetic and you compute:

- ✔ roughly six billion people on earth;
- ✔ roughly one billion kids to get presents;
- ✔ 24 hours to deliver all those presents;
- ✔ 1,000,000,000 ÷ 24 = 41,666,666 kids per hour;
- ✔ 41,666,666 ÷ 60 = 694,444 kids per minute;
- ✔ 694,444 ÷ 60 = 11,574 kids per second;

that you start to think that something goofy is going on.

Some things are hard to believe. In the previous chapter we told you that almost all of the non-water weight of giant trees doesn't come from the dirt they are planted in, but from the air.

I, your reader, think that you are
>> **telling me a tall tale,**
>> **pulling my leg,**
>> **"funning" me,**
>> **laying it on thick, and**
>> **fibbing.**

I'm not some little four-year-old kid that will believe any hokum* you say. You did the arithmetic on Santa Claus, and now I'll do the arithmetic on your theory that most of the non-water weight of plants comes from the air.

Fact number one: Air is a mixture of gases. By weight, it's 75.54% nitrogen, 23.14% oxygen, and 1.27% argon. That adds up to 99.95% of the gases in the air.

∗ Hokum is silly nonsense.

Chapter Sixteen   *Out of Thin Air*

Fact number two: Anybody who has ever studied the chemistry of biological things (that's called organic chemistry) knows that the chemical element carbon is the basis of all the organic compounds.

Trees and bees and fleas are all composed of carbon-based compounds.

All I said was that *plants* get most of their non-water mass from the air. Animals get their carbon by munching on plants (or other animals).

But did you read fact number one? There's no carbon in the air. Nitrogen, oxygen, and argon are 99.95% of the atmosphere.

What's the fourth most common gas in the atmosphere?

Get serious, Mr. Author. Nitrogen, oxygen, and argon are the big three. The fourth, fifth, sixth, and seventh most common gases are a trivial 0.05% of the atmosphere. They are not worth talking about.

You are avoiding the question. What is the fourth most common gas in the atmosphere?

It is carbon dioxide—CO_2 if you like its chemical formula. And now you are going to tell me that fields of wheat, forests of trees, and meadows of flowers get their carbon mass from this trace of CO_2 in the air? You know math: 0.05% is the same as 0.0005, which is $\frac{5}{10000}$ which reduces down to $\frac{1}{2000}$. Do you know how small that is? Here are 2000 dots. And you are telling me that plants can sift through that many molecules of gas and pick out the one that is CO_2 and use it to support the biology of the whole earth?

✧✧
✧✧
✧✧
✧✧
✧✧
✧✧
✧✧
✧✧
✧✧
✧✧
✧✧
✧✧
✧✧
✧✧
✧✧
✧✧
✧✧
✧✧

Chapter Sixteen   Out of Thin Air

Yes.

Yeh. Sure. Ha! Ha! Ha! Ha! Ha! Ha! Ha! Ha! Ha! Ha! Ha! Ha! And next you are going to tell me that in geometry the parallel postulate isn't true.

The Parallel Postulate seems really obvious, but it isn't necessarily true. If you have a line ℓ and a point P, you can have many lines through P that are parallel to line ℓ. I explain that in Chapter 11½ of *Life of Fred: Geometry*.

And you are going to tell me that there are numbers that are neither positive, negative nor zero.

Yup. We do that in algebra. And we also talk about 2^{-3} and $8^{1/3}$.

Education is supposed to be an introduction to reality. Reality contains many mind-blowing surprises. Education can't be boring—unless, of course, it is done poorly.

Now, let me get back to describing how air makes trees. Oops, we're at the end of the chapter.

Chapter Sixteen   Out of Thin Air

Your Turn to Play

1. Please check my work and show that 75.54% nitrogen, 23.14% oxygen, and 1.27% argon add up to 99.95%.

2. 99.95% = $\frac{9995}{10000}$ Reduce this fraction.

3. The union of {1, 5, 7} and {1, 2, 7} is sometimes written as {1, 5, 7} ∪ {1, 2, 7}. Find {1, 5, 7} ∪ {1, 2, 7}.

4. Let T be the set of all trees.
   Let P be the set of all plants.
   Then T is a **subset** of P. Every tree is a plant.
   (Official definition: T is a subset of P if every element of T is an element of P.)
   Is P a subset of T?

5. Name some other set that is a subset of P.

· · · · · · · **COMPLETE SOLUTIONS** · · · · · · ·

1. 75.54%
    23.14%
  + 1.27%
    99.95%

2. To reduce $\frac{9995}{10000}$ we have to find a number that divides evenly into both 9995 and into 10000. 2 doesn't work since 9995 isn't even. 3 doesn't work. 5 does work.

   Dividing top and bottom by 5, $\frac{9995}{10000}$ reduces to $\frac{1999}{2000}$

3. {1, 5, 7} ∪ {1, 2, 7} = {1, 2, 5, 7}. One "spelling rule" is that you don't list the elements of a set more than once.
   E.g., you don't write {1, 1, 2, 5, 7, 7}.

4. P is not a subset of T since not every plant is a tree.

5. Answers will vary. The set of all radishes is a subset of the set of all plants. The set of all plants in my backyard is a subset of P.

Chapter Seventeen
Photosynthesis

Fred gave Glenda $2 for her plant. Fred couldn't put the four-foot plant in his paper bag, so he just held it in his hand.

$$\begin{array}{r} 74999.31 \\ -2.00 \\ \hline 74997.31 \end{array}$$

"I'm still amazed to think that this plant got most of its non-water weight from the air," Fred said. "Didn't it get anything from the dirt it was planted in?"

"Sure," said Glenda. "It got water through its roots, and it got various tiny amounts of chemicals from the soil. It is a little like when you take a vitamin pill."

She pulled Prof. Eldwood's *Heart of Biology* book off the shelf.

Chapter One
Making Plants Out of Air

There is about 0.05% carbon dioxide in the atmosphere. That's all it takes. Plants are very patient. They keep sorting through all the nitrogen, oxygen, and argon until they find that very special CO_2 molecule. They invite the carbon dioxide inside their leaves and introduce her to Mr. Chlorophyll. (The green stuff in plants is called chlorophyll.) The chlorophyll has been sunning himself and is all full of energy from the sunlight. Together, the CO_2 and the chlorophyll sit down and have a sip of water.

And together they make sugar.

The End.

Chapter Seventeen Photosynthesis

"That's an incredibly short book," Fred said. "I get the feeling that he may have left out a couple of the details."

"How complicated do you want to get?" Glenda asked. "It's just carbon dioxide and water combining in the presence of sun-energized chlorophyll." Glenda drew a picture of the process.

Fred thought that picturing sunlight energy as a body-builder duck was cute, but he felt like he was being treated like a five-year-old. He said, "Could you give me the version for twelve-year-olds?"

Glenda wrote . . .

carbon dioxide + water + energy from sunlight stored in chlorophyll   gives sugar + oxygen

This wasn't enough for Fred. He asked for the version for 15-year-olds, and Glenda wrote . . .

$$6CO_2 + 6H_2O + \text{energy} \rightarrow C_6H_{12}O_6 + 6O_2$$

Glenda explained, "This means that six molecules of carbon dioxide plus six molecules of water in the presence of energy will give you one molecule of sugar and six molecules of oxygen. This whole process is called **photosynthesis**. *Photo* means "light" and *synthesis* means to assemble things together." She thought that this explanation would satisfy six-year-old Fred. She was wrong.

Fred asked for the version for 18-year-old college students.

Chapter Seventeen   Photosynthesis

Glenda continued, "It takes the energy stored in the chlorophyll to make this reaction happen. Carbon dioxide and water don't naturally want to convert into sugar and oxygen. The extra energy, which is supplied by sunlight, is stored in the sugar molecule."

Fred giggled to himself. This was fun. "How about the version for 24-year-olds?"

Glenda couldn't believe that this five-year-old was asking for more information about photosynthesis. She told him that the sugar produced by photosynthesis is what makes oranges taste sweet. But there is a limit to how much sugar a plant can handle since sugar dissolves in water. Too much sugar could interfere with the health of the plant. So photosynthesis continues by converting sugars into starches (think potatoes) and oils (think corn oil). Starches and oils don't dissolve in water.

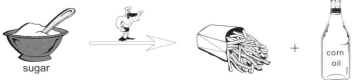

"Most people," she continued, "don't know that photosynthesis doesn't stop with making sugar. It makes the wood in trees. And adding more sunlight energy gives you oil, which has more energy stored in it (calories) than sugar or starch."

Before Fred could ask for the 30-year-old version, Glenda took a french fry and put it where it would do the most good. With his mouth full, she figured that he couldn't ask any more questions.

Your Turn to Play

1. Each year plants convert 600 billion tons of CO_2 into sugars, starches, and oils. Using conversion factors (1 ton = 2000 pounds), how many pounds of carbon dioxide are removed from the atmosphere each year?

2. Fred took the french fry out of his mouth and put it in his pocket "for later." His pockets contained a lot of food that he had never gotten around

to eating. Often, when he was sitting at his desk in his office, he would transfer the food from his pockets to his desk drawers. The drawers never became full. Why? The ants around KITTENS University are very smart. When they are not playing chess, they are out looking for food left in desk drawers.*

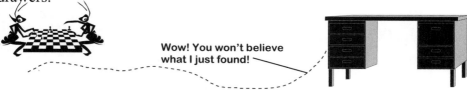

There were 8 grams of ants working on Fred's food drawer. How many ants was that? (783 ants weigh one gram.)

3. This is the set of the two most important chess pieces: {King, Queen}. Name all four possible subsets of that set.

. . . . COMPLETE SOLUTIONS . . . .

1. $\dfrac{600{,}000{,}000{,}000 \text{ tons}}{\text{year}} \times \dfrac{2000 \text{ lbs.}}{1 \text{ ton}}$

= 1,200,000,000,000,000 lbs per year

= 1200 trillion lbs. per year

= 1.2 quadrillion lbs. per year

one thousand	1,000
one million	1,000,000
billion	1,000,000,000
trillion	10^{12}
quadrillion	10^{15}

2. $\dfrac{8 \text{ grams}}{1} \times \dfrac{783 \text{ ants}}{1 \text{ gram}} = 6264$ ants

3. {King, Queen}, {King}, {Queen}, and { }.

★ You think I'm kidding about "when ants are not playing chess, they are out looking for food"? I can prove it. Think of all the times you have seen ants out looking for food. Were they playing chess? No they weren't. That shows that only after they finished their chess game did they go out looking for food.

   Do you like my reasoning?

   There is a subfield of mathematics called logic that analyzes arguments to see which ones are valid and which ones are nonsense.

Chapter Eighteen
Fred the Heterotroph

Fred rubbed his eyes. It was like a black cloud had just passed overhead. Everything was much darker. He didn't feel faint, but it was like he was running out of energy. He reached into his pocket and broke off a small piece of the french fry.

He placed it near his oral cavity (mouth) and used his **incisors** to bite off a tiny piece of the fry. (The incisors are the cutting teeth in the front of the mouth. There are four on the top and four on the bottom.) His tongue then moved the fry back to the flat grinding teeth (the **molars**).

After several minutes of chewing, several things happened:
1. The **cell*** walls in the food were broken open.
2. The food got juicy. Fred's spit (**saliva**) did several things. It contains an enzyme (salivary amylase) that breaks down the starch in the potato into sugar. The **salivary amylase** [SAL-eh-very AM-eh-lace] doesn't do anything to meat or to candy (sugar). It just works on the starch. Saliva [sal-LIE-veh] also makes things easier to swallow. Corn chips can be rough on the throat if they are swallowed when they are still dry.
3. Finally, after several minutes of chewing, Fred realized that you were supposed to do something. He swallowed.

The rest of **digestion** happens automatically.

First it passes the **pharynx** (the tube that leads to both the lungs and the rest of the digestive system). Here is where we decide whether it is air or food that the oral cavity is sending. If you swallow a piece of gum (not recommended), you sure don't want it bouncing around in your lungs.

Then the food goes down the **esophagus** [eh-SOF-ah-gus] to the tummy. Of course, if you are standing on your head when you eat, the food will travel *up* the esophagus to your tummy. It is not gravity that moves the food to your tummy.

---

* Cells are little "boxes" that you can see under a microscope when you look at the tissue of living things. They look really simple when you first see them, but when you learn about meiosis and mitosis, and the fact that eukaryotic cells contain many membrane-bound organelles, which prokaryotic cells don't have, etc., you find that cells are really complicated. There are lots of things that scientists haven't yet figured out about cells.

Then the tummy (stomach).
Then the small intestine.
Then the large intestine.
Then it's time to go to the bathroom.

We have left out a great number of details, such as how the stomach, pancreas, liver, and gallbladder dump about 9 liters* of fluid into the first part of the small intestine each day. Happily, about 95% of that fluid is reabsorbed back into the body through the lining of the small intestine.

For more information, check out the apocryphal 837-page book by Prof. Eldwood entitled *Digestion—Life Along the Alimentary Canal*, 1863. (Alimentary canal = gastrointestinal tract) [al-eh-MEN-teh-ree]

That tiny bit of french fry that Fred swallowed was partially digested by the salivary amylase in his mouth. It met the gastric juices in his stomach. Then on to his small intestine. It seems silly to call it the *small* intestine. In adults it is 6–7 meters** long. Maybe they call it the small intestine because it is only about an inch in diameter.

People who don't like small intestines sometimes label them as the PLACE OF FINAL DIGESTION. That sounds so negative. It is true that his french fry gets hit with a bunch of enzymes from the intestinal walls, enzymes that the pancreas dumps into the first part of the small intestine, and bile salts that are produced by the liver. All of these complete the job of breaking down the potato + grease + salt (= french fry) into its digestible components.

But I like my small intestine. I call it the Welcome Center. It is here that the broken down components leave the tunnel we call the alimentary canal and enter into the blood stream.

These happy little sugars and amino acids take a boat ride around the body until they hop off and enter the cells.

You'll never guess what happens next. I think it is weird. You get sunlight inside your cells.

---

* A liter is like a big quart. 1 liter = 1.057 quarts

** A meter is like a big yard. 1 meter = 1.094 yards

Chapter Eighteen   Fred the Heterotroph

You remember how plants (which are called **autotrophs**) take water and carbon dioxide, add some energy from sunlight and produce sugars (and starches and oils).

Autotrophs take inorganic chemicals (water and parts of the atmosphere) and turn them into organic chemicals. It takes energy to do this. A favorite saying among autotrophs is, "Make sugar while the sun shines." Only about 20% of all living things can do this trick.

The rest of us (100% − 20% = 80%) are **heterotrophs**. We work in reverse. We eat the sugars, starches, and oils. We breathe in the oxygen. And inside our cells we do the opposite of what the autotrophs (plants) do. We heterotrophs produce carbon dioxide, water, and energy.

Wait a minute! I, your reader, think you are a bit mixed up. You wrote on the previous page, "You get sunlight inside your cells." Those are your very own words. When I go into a dark room and look at myself, I do not glow. You had better explain yourself.

I want to talk about this "reverse photosynthesis," and you want to discuss my veracity.*

This requires an intermission.

Intermission

There is more than one way to express a truth. When you see a beautiful sunset (or a beautiful pizza or a beautiful geometry proof), you could just type `It is pleasing to the pleasure centers of my brain.` Or `My respiration rate increases.` Or `My eyes dilate.`

These are all factually, literally true. But they feel like they were written by a robot.

---

* Veracity is the truthfulness of a person. [veh-RAS-it-tee where RAS rhymes with MASS]

97

Chapter Eighteen   Fred the Heterotroph

> What if I write **When I saw that sunset, pizza, or geometry proof, my heart skipped a beat.** Or **The world burst forth into song.** These are not literally true. My heart did not develop an irregular rhythm. Everyone wasn't running around singing. But in some real sense, these statements are more true than just stating the facts about my breathing rate.
>   That is why we read good poetry and great novels—they tell more truth than facts like . . .
>   1 cubic yard = 807.9 quarts
>   Washington was the first U.S. president.
>   Yosemite is spelled Y-o-s-e-m-i-t-e.

When I sing to my wife, "You are my sunshine," she knows that I don't mean that she is composed of the photons from the sun.

When I report we have sunlight in our cells, what does this mean? It means that the energy,

which was sunlight,

which, using chlorophyll, was stored by autotrophs as sugars, starches and oils,

which was eaten by heterotrophs like us,

which was digested in the oral cavity, the stomach, and the small intestine,

which was ferried by the blood stream to the cells,

was there released as energy.

What started as sunlight energy is now energy in our cells. It's true that we don't glow. But we dance, we sing, we grow, we give warm hugs.

Your Turn to Play

1. Physicists like to talk about all the different forms that energy takes. The energy in sunlight is turned into the chemical energy in food, which the cells turn into heat, into electrical impulses in our nerves, into motion in our muscles, and into. . . . If the cells of your body can't use up all the

energy they receive, some of it is put into long-term storage for use when food may become scarce. What's this storage called?

2. The different parts of our bodies require different amounts of energy. One way to measure the amount of energy used is to measure the amount of oxygen used. For example, our bones use a lot less oxygen than our muscles use. There is one part of our body that uses a disproportionate share of the oxygen we breath in. It weighs about three pounds and uses one-fourth of the oxygen supplies in our body. What part of the body is that?

3. What is one-fourth as a percent?

4. Suppose you have a 3-pound brain in a 120-pound body. What percent of your body weight is brain? In the traditional formulation, this would be expressed as, "3 is what percent of 120?"

5. If you weigh 120 pounds and 30% of that is fat, how many pounds of fat are you carrying around?

. . . . COMPLETE SOLUTIONS . . . .

1. I call it my extra sunshine. Everyone else calls it fat.

2. If you guessed "my big toe," you get a grade of NC (Not Close). If you guessed "my heart," you get a grade of NTBW (Nice Try But Wrong). The energy hog is the brain. If you were to cut off the blood supply to the brain, it would use up all its oxygen in TEN SECONDS. This is why your mother told you two things: (1) Remember to take your lunch and (2) Keep breathing.

3. $\frac{1}{4}$ = 4)$\overline{1.00}$ (= .25) = 0.25 = 25%   The brain uses 25% of the oxygen we breath in.

4. 3 = ?% of 120. The rule that was in *Life of Fred: Decimals and Percents* was If you know both sides of the of, then multiply. Otherwise, divide the number closest to the of into the other number.
So we divide 120 into 3.   120)$\overline{3.000}$ (= .025)   So your brain is 2.5% of the weight of your body.

5. 30% of 120 = 0.3 × 120 = 36 pounds. Imagine climbing stairs wearing a backpack with seven 5-pound sacks of sugar in it.

Chapter Nineteen
Eyes, etc.

Even after Fred's giant meal of one gram* of french fries, everything still seemed to be getting darker. Fred now weighed 37 pounds plus one gram. What's happening? he thought to himself. Am I going blind? Four years ago, when he was only one year old, he had realized that most people have eyeballs, and he had only eyedots.

  Fred had looked at science books to find out what normal eyes were like. The first book he looked at was entitled Mommy Explains. It began, "Eyes are used for seeing." Too babyish. He put the book back on the shelf.

The second book was entitled *Transducing Energy in Heterotrophs*. The first chapter dealt with **thermoreceptors**. It described how pit viper snakes have heat-detecting organs in their heads. Even in total darkness, they can sense the body heat of a mouse up to a meter away. The chapter concluded with a discussion of the thermoreceptors in the **hypothalamus** in human brains that check how close to 98.6° F the body is. Then the hypothalamus tells the body to either shiver or sweat.

The second chapter was about **chemoreceptors**, which detect chemicals in the environment. In this fancy book, they said that in humans we have chemoreceptors for **olfaction** and **gustation**. Fred looked up these words in the dictionary. They mean *smell* and *taste*. Fred couldn't figure out why the author was making simple things hard. Is he just trying to impress me with how smart he is? Fred thought. That's dumb.

The third chapter discussed fish that live in deep dark waters or in muddy waters or turbulent waters where sight and smell don't work well. They use **electroreceptors**—sensing electric fields—to catch prey.

The fourth chapter was about **audioreceptors** (= hearing). This was a frightening chapter for Fred. He realized that besides having eyedots, he lacked pinnas. He thought about it for a while and decided that ears were funny-looking. He was glad he didn't have any.

Fred with a pinna

★ One gram is about the weight of a raisin.

Chapter Nineteen   Eyes, etc.

The fifth chapter discussed **mechanoreceptors**. That was the sense of touch. The mechanoreceptors are concentrated in the tongue, the lips, the face, and the fingertips. That's why it's not a good idea to bang your lips with a hammer.

Chapter six talked about **proprioceptors**, which are the mechanoreceptors in joints and muscles. Without even looking, you can use your proprioceptors to tell whether your hand is open or shut.

In chapter seven Fred hit pay dirt. It began, "**Photoreceptors** transduce light energy." Fred ran back to his dictionary and looked up *transduce*. It said, "*Transduce* is to change energy from one form into another."

Our nerves transmit electrical signals, so all these "receptors" transduce the stimuli (heat, odors, sound, pressure, light) into electrical energy.

The book even had a picture of a normal human eye. On the back wall of the eye are the photoreceptors called the **retina**. [RET-in-ah] The retina transduce light into electrical signals which are sent along the optic nerve to the brain.

"What's happening?" Glenda asked. "It's getting dark."

Fred realized that his photoreceptors were not malfunctioning. It really was getting darker. The lights were going out.

New Mall

New Mall without lights

"How wonderful!" Fred exclaimed.
"Wonderful?" Glenda asked.
"I'm not going blind," Fred answered.

Chapter Nineteen   Eyes, etc.

Your Turn to Play

1. In a group of 400 people, if the lights went out, two percent of them might exclaim, "Wonderful!" How many people is that?

2. Fred really meant it when he said, "Wonderful!" He was happy not to be going blind.

    The other seven people would be speaking ironically.*

    Fred is what percent of the 400?

3. While reading that difficult book, *Transducing Energy in Heterotrophs*, Fred sweated off a gram. He then weighed exactly 37 pounds. How many kilograms is that? (One pound is approximately 0.45 kilograms.)

. . . . COMPLETE SOLUTIONS . . . .

1. We are asking what 2% of 400 equals. From the rule given three pages ago: *If you know both sides of the* of, *then multiply.*

2% × 400 = 0.02 × 400 = 8 people would exclaim "Wonderful!"

2. We are asking 1 = ?% of 400. From the rule given three pages ago: *If you don't know both sides of the* of, *divide the number closest to the* of *into the other number.*

$$400 \overline{\smash{)}1.0000}^{.0025} = 0.25\%$$

3. Using the conversion factor approach:

$$\frac{37 \text{ lbs.}}{1} \times \frac{0.45 \text{ kilograms}}{1 \text{ lb.}} = 16.65 \text{ kgs.}$$

――――――――――

* To speak ironically is to use **irony**. (I-ron-knee) Irony is saying the exact opposite of what you mean. Suppose you are standing at the bus stop. It's raining hard, and you are freezing and getting soaked to skin. You look down the street and notice that a tornado has just lifted your bus up into the sky. You turn to the guy next to you and say, "Beautiful weather, isn't it?"

Chapter Twenty
Negative Numbers

Coalback appeared in the doorway to Glenda's shop. He shined his flashlight* on Glenda and announced, "I've come for today's rent—$400."

"So far today," Glenda told C.C. Coalback, "I've sold a two-dollar plant. I don't have $400, but you are welcome to all the money I've made."

Coalback turned to Fred. "Hey kid, what's that you got in your hand?"

Fred held up the four-foot plant that he had just purchased from Glenda.

"No. What's in your other hand."

Before Fred could say that it was a paper bag with a shovel and $74,997 in it, C.C. Coalback reached in and took $800.

$$\begin{array}{r} 74997.31 \\ -800.00 \\ \hline 74197.31 \end{array}$$

As he left, he said, "That will pay for today's and tomorrow's rent. You were trying to cheat me by hiding your cash on your kid."

"But," said Glenda, "this little man is a customer. He's not my child." By this time, Coalback was out of earshot. He had headed next door to Ferdinand's Farming to collect the rent.

Stunned, Glenda and Fred stood there in the darkness. Minutes passed.

Meanwhile, Coalback had visited store after store: Ferdinand's Farming, Eve's Garden, Dakota's Digging Equipment, Cameron's Canning Supplies, and Beatrice's Better Blooms. None of these had had any customers. None of them could pay the rent.

---

* In Britain they don't call it a flashlight. They call it a torch. Some of them carry a torch in the boot of their automobile (flashlight in the car trunk).

Chapter Twenty   Negative Numbers

When Coalback got to Abby's Agricultural Supplies, he told Abby, "You have a very neat store. Everything is organized well. Have you had any customers? I've come for the rent."

"Oh yes. I had a boy come in and buy a sand pail shovel for 69 cents."

"Is that all you got? You owe $400."

Abby explained that the boy had paid by check. She asked Coalback, "Do you have change for this check?"

"Ha! That check isn't worth kitty litter," Coalback grinned. "It's not worth a nickel. I closed that account about ten minutes ago."

Coalback had lied when he had told the store owners that there would be "an uncountable number of customers" and no one would have "trouble paying the rent of $400." He had rented out the 80 stores by promising the tenants things that weren't true. He hadn't even paid the electric bill because he knew that the mall would never be a success. He was practicing what he called the Three G's.*

It was time for some more grabbing.

Coalback headed back to Glenda's Garden Goodies. He shined his flashlight in Fred's face and asked, "Son, have you ever thought of owning a mall?"

"But sir," Fred said, "a mall with 80 stores in it must cost millions of dollars. I only have $74,197 in this bag and 31 cents in my pocket."

"Keep the change, kid." Coalback snatched the bag of money and handed Fred the deed to the mall. On his way to his car, Coalback looked in the bag, took out the plastic shovel and tossed it on the sidewalk.

The deed was real. It was valid. It wasn't like the $75,000 bum check that Fred had received as contest winner. Fred really did own a mall with 80 stores in it. Not bad for a five-and-a-half year old.

---

* 3 G's = Gab, Grab, and Go. You lie (gab). You take all the money you can get your hands on (grab). And you leave town (go).

Chapter Twenty — Negative Numbers

Your Turn to Play

1. Each store cost $62,500 to build. (That includes the cost of sidewalk in front of the store.) How much is Fred's mall worth? (There are 80 stores.)

2. Fred gave up $74,197 for a five million dollar mall. (We can't say he *paid* $74,197, since the money was snatched out of his hands.) How much did Fred gain in the transaction?

3. There was an electric bill outstanding that Fred had to pay to get the lights on in the mall. $8,378.
There were the property taxes that were overdue. $3,209.
There was the sidewalk contractor, the carpenters, the electricians, the plumbers, the surveyors, and the painters who needed to be paid for the construction of the mall. These workers were owed $914,216.
   What is the total of these bills?

4. Starting with a mall that was worth five million dollars and subtracting Fred's payment of $74,197 and the outstanding bills of $925,803, Fred still had a net gain of $4,000,000. If this is correct, Fred thought to himself, my Sunday school offering will be $400,000.
   Fred had forgotten one little tiny (irony alert!) item. He had started with a five million dollar property. He had subtracted his payment. He had subtracted the construction costs that were still owed. What else could affect Fred's four million dollar gain?

......COMPLETE SOLUTIONS.......

1. Most people would just multiply: $80 \times 62{,}500 = \$5{,}000{,}000$, but this can also be done with conversion factors.

$$\frac{80 \text{ stores}}{1} \times \frac{\$62{,}500}{1 \text{ store}} = \frac{80 \:\cancel{\text{stores}}}{1} \times \frac{\$62{,}500}{1 \:\cancel{\text{store}}} = \$5{,}000{,}000.$$

2. $\$5{,}000{,}000 - \$74{,}197 = \$4{,}925{,}803$ he gained in the transaction.

3. $\$8{,}378 + \$3{,}209 + \$914{,}216 = \$925{,}803$.

4. A lien (pronounced *lean*) on your property is a legal right for someone to seize or sell your property if you don't pay a claim or debt. When you get a bank loan, the bank has a lien on your property. Coalback had gotten

a loan from Kittens Bank for $4,800,000. If you subtract $4,800,000 from Fred's $4,000,000, you get a *loss* of $800,000.

$4,000,000
− $4,800,000
loss of $800,000

Fred got out a piece of paper and wrote it all down . . .

My Gain	My Losses	
I got a property worth 5,000,000	I paid	74,197
	I owe the workers	925,803
	I owe Kittens Bank	4,800,000
	total	5,800,000

In accounting, they write $5,000,000 − $5,800,000 = $<800,000>
In math, 5,000,000 − 5,800,000 = −800,000.

−800,000 is a **negative number**. It is a number less than zero.

One meaning of the word *dead* is "absolutely." If you say you are dead tired, that means you are completely tired.

If you are dead broke, you have zero net worth—what you owe matches what you own.

Fred was worse than dead broke. His net worth was −$800,000. If someone gave him $800,000, then he would be dead broke. Here's the picture:

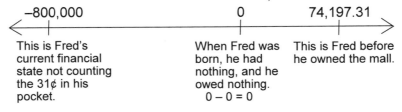

−800,000	0	74,197.31
This is Fred's current financial state not counting the 31¢ in his pocket.	When Fred was born, he had nothing, and he owed nothing. 0 − 0 = 0	This is Fred before he owned the mall.

Chapter Twenty-one
Eyelashes

In the beginning, Fred had just wanted to buy some gardening supplies so that he could plant his victory garden. He now held a four-foot plant in one hand and the deed to an 80-store mall in the other.

He said goodbye to Glenda and left her store. He saw his plastic shovel on the sidewalk and picked it up.

Two-year-old Sally walked up to him and said, "Hello mister man. Have you seen my brother Jimmy or my mommy? I think I lost them."

Fred didn't know what to say. He thought to himself *When I grow up and get married, I want to have a lot of kids. Maybe six or maybe twelve. And I would make sure never to lose any of them.* He said to Sally, "Let's sit down and think about it." They sat on the outdoor bench between Glenda's and Ferdinand's.

"Can I play with your shovel?" she asked. She took it out of Fred's hand and waved it around. Then she pretended to dig imaginary dirt off the bench. After a minute, she got bored.* "Do you gotta crayon?" she asked.

Fred handed her his pencil. He got off the bench and looked up and down the street for Jimmy and Sally's mother. Sally pulled a piece of paper out from under the plant that Fred had placed on the bench. She turned it over to the blank side and began to draw.

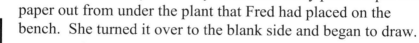

When Fred came back, she showed him her art work. "It's a picture of me when I go to school someday."

---

* One effect of having read a lot of good books (= a good education) is that you don't get bored easily. Your mind is filled with a lot of interesting ideas that need to be thought about—a lot of questions that have no easy answers. There's plenty to think about.

You will meet people who need entertainment every moment of the day. Nowadays, many waiting rooms (in auto repair shops, in doctor's offices) have a television on. Text messages read, "How R U? I haven't heard from you in an hour." Cell phones help fill in the "boring" three minute walks between classes.

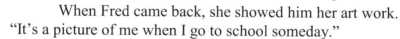

Deprived of outside stimulation for more than a minute, acute discomfort sets in.

Chapter Twenty-one   Eyelashes

She erased one eyedot and announced, "And here is a picture of me winking. When I go to school, everyone is going to like me because I can wink at them."

"There she is!" said Jimmy.

Pulling Jimmy along, Sally's mom walked over to her. "How many times have I told you not to go wandering off by yourself!"

"I'm not alone," said Sally. "I'm with that funny-looking guy with the big nose."

She yanked Sally off the bench and marched off with her two kids. Sally looked back at Fred and winked. And she waved. She had her picture in her hand. Sally now owned the mall.

When Sally had been younger,* she used to stand in front of the mirror and practice winking. When she first tried to wink, both eyes would shut at the same time. Eventually, she learned to shut them one at a time.

Then she discovered that she had eyelashes. She called them "eye fur" until her mother corrected her.

Sally had long eyelashes. Her mother and her brother Jimmy both had short eyelashes. Long eyelashes made winking more fun. Her mother was really jealous of Sally's eyelashes. She used to grumble when she applied mascara to her short eyelashes to make them look as long as her daughter's.**

Eyelashes come in only two lengths—short and long. This makes it a very easy trait to look at from a biological point of view. There are only two versions of the **gene** that controls eyelash length.

With many human traits things are more complicated. There are multiple genes that affect intelligence, skin color, and chances of

---

* When Sally was $1\frac{1}{2}$. That is less than her current age of 2. In symbols we write $1\frac{1}{2} < 2$. The little pointy end of $<$ points to the smaller number.
These are true statements: $3 < 3{,}000{,}000$
$$\pi < 4$$
$-800{,}000 < 0$
$0 < 74{,}197.31$

** This is, of course, a violation of the Tenth Commandment, which says that you are not to go nuts over anything that someone else owns.

Chapter Twenty-one   Eyelashes

developing various health problems. These traits are called **polygenic**. Those polygenic traits are definitely not the place to begin the study of genes. When you first were learning to sing, you started with "Mary Had a Little Lamb," not with opera.

The study of genes—called **genetics**—really started happening in 1900. That's when Hugo deVries (Netherlands), Carl Correns (Germany), and Erich Tschermak (Austria)—each working independently—started to piece together the way that genes worked.*

Except for Mendel, virtually all biologists before deVries, Correns, and Tschermak thought that the traits of kids were just some kind of mixture of the traits of the parents. Like mixing paints. Like making salad dressing from vinegar and oil.

> In the many centuries before 1600, almost everyone thought plants ate dirt.
>
> In the many centuries before 1900, almost everyone thought that the traits of parents were blended in their kids.

Here's what we know about eyelash length today. Everybody has two genes that determine eyelash length. They got one of their genes from their dad and one from their mom. Let's call those genes S (short) and L (long).

So everybody is either   SS

or SL

or LL when it comes to eyelash length.

There are only three possible **genotypes**—SS, SL, or LL.

If your genotype is SS, you will have short eyelashes. (Like Jimmy and his mom.)

If your genotype is LL, you will have long eyelashes.

---

* We have got to give credit to a monk named Gregor Mendel. For two years he worked with garden peas and figured all the genetics that deVries, Correns, and Tschermak were to rediscover about 35 years later.

In 1866 Mendel published his work . . . and was ignored.

It's the Double R Rule: *It happens when you are Right at the Right time.*

Mendel was right, but the world wasn't ready for him.

Chapter Twenty-one Eyelashes

If your genotype is SL, you won't have medium length lashes. You will have long eyelashes. (No blending for traits controlled by a single gene. Please check your calendar—we are not living before 1900.)

The L gene is the **dominant gene** (over the S gene). It wins.

Your Turn to Play

1. Jimmy and his mom have short eyelashes. What is their genotype?
2. Sally has long eyelashes. Is it possible to say what her genotype is?
3. Do we know the genotype of Sally's father?
4. Does Sally's father have short or long eyelashes?
5. If you know a person's phenotype (by looking at him), can you determine his genotype?

.......**COMPLETE SOLUTIONS**.......

1. They couldn't be SL or LL. They both must be SS.
2. When students are first learning about genetics, they usually say that Sally could be either SL or LL, since either of those will give Sally long eyelashes. But Sally couldn't be LL. Recall, you get one gene from your father and one from your mother. From the previous question, we know that Sally's mom is SS. So Sally must have received an S from her mom. So Sally's genotype must be SL.
3. We know that Sally got an L from her father, since her genotype is SL, and she got her S from her mother. Her father must either be SL or LL. But wait! Son Jimmy is SS (from question 1). He must have gotten an S from each of his parents. The father must be SL.
4. From the previous question, we know he is SL. L dominates S, so the father must have long eyelashes.

His genotype is SL.

His **phenotype** is long eyelashes. The phenotype is the observable trait. Your genotype determines your phenotype.

5. You see someone with long eyelashes (phenotype). They could be either SL or LL (genotype). No, you can't always tell. (You can tell in some cases. For example, see question 1.)

Chapter Twenty-two
Variation

Fred never spent much time looking in a mirror. It was so discouraging. There was no hair to comb. No ears to look at. He didn't even have eyelashes. Once he thought to himself *I look like a duck—no hair, no ears, no eyelashes.* And besides, when he got too close to the mirror, his nose would scratch the glass.

Eyelashes?

He thought about the variation that he noticed in humans. If you were the lucky owner of eyelashes, would you rather have them long or short? It depends. If you lived in some hot, humid country that had lots of insects, long eyelashes would be great to bat the insects away from your eyes. If you lived in some snowy spot, having long eyelashes that got icy would be horrible.

Tall and skinny people tend to feel cold more easily. They lose body heat more quickly than short and round people. Living and working in snowy climates, you are at an advantage if you are not tall and skinny.

For many phenotypes (traits that you can see), one variation is not always better than another. It often depends on what environment you are in.

Fred is very short (36 inches) for a 5½-year-old boy. One reason is that he never seems to be very hungry. That one gram of french fries that Fred ate eleven pages ago will keep him full for days. In what environment could this be an advantage?

But Fred wasn't in front of a mirror. And he wasn't in a concentration camp where food is scarce. He was just your average (irony alert) boy walking down the street carrying a plant in one hand and a plastic shovel in the other. And he was playing with numbers in his head.

The square of 7 is 49. $7 \times 7 = 49$. $7^2 = 49$.
The square of 15 is 225. $15^2 = 225$.
The square of 1.5 is 2.25. $1.5^2 = 2.25$.

Chapter Twenty-two Variation

What number squared is equal to 64? $x^2 = 64$. He let "x" be the unknown he was looking for. He was looking for a number that made $x^2 = 64$ true. How about x = 8?

Then he thought of the equation $x + x = 2x$. What numbers make that true? It's true for x = 6 since $6 + 6 = 2×6$.
It's true for x = 11 since $11 + 11 = 2×11$.
It's true for x = 0 since $0 + 0 = 2×0$.
Every number makes $x + x = 2x$ true.

> ### Intermission
>
> Using x for some unknown number is the beginning of algebra. When you write 2x = 18, and you figure out that x must be 9, you are doing algebra.
>
> When you were just a kid, you learned your multiplication tables and would write 3 × 4 = 12.
>
> In algebra, when we want to write *two times x*, we write 2x and not 2 × x. Can you see why? The times sign can get mixed up with the x. We don't use the times sign in algebra.
>
> In algebra, when we want to write 3 × 4, we write it as 3(4) or (3)(4).

Fred thought of the equation $x + 1 = x$. What number would make that true?
He tried x = 4. $4 + 1 \stackrel{?}{=} 4$.   No that doesn't work.
He tried x = 7. $7 + 1 \stackrel{?}{=} 7$.   No that doesn't work.
He tried x = π. $π + 1 \stackrel{?}{=} π$.   No that doesn't work.
He tried x = 0. $0 + 1 \stackrel{?}{=} 0$.   No that doesn't work.

Fred couldn't find any numbers that made $x + 1 = x$ true. That equation has no solution.

Just as there is variation in humans, there is variation in equations. Some have one solution. $x + 5 = 9$ has only one solution: x = 4.
Some equations have no solution: $x = x + 6$.
Some equations are true for every value of x: x = x.

Chapter Twenty-two Variation

Your Turn to Play

1. Suppose one parent had the genotype LL (where L stands for Long eyelashes), and one parent had the genotype SS (where S stands for Short eyelashes). L is dominant over S.

   Suppose they had 400 kids. What percentage of them would have long eyelashes?

2. Suppose both parents had the genotype LS. Roughly, what percentage of their children would have short eyelashes?

3. S is a **recessive** gene. It is dominated by L. L masks the effect of S in the phenotype.

   True or False: *It is always possible to tell the genotype if the phenotype is that of a recessive gene.*

4. Find the two numbers which make the equation $x^2 = x$ true.

5. Find the two numbers which make the equation $x^2 = 3x$ true.

6. What number makes $x + 55 = 77$ true?

. . . . COMPLETE SOLUTIONS . . . .

1. Every one of the kids would have the genotype LS. Since L is a dominant gene, 100% of the kids would have long eyelashes.

2. You get one gene from each parent. There are four equally likely cases:
   - L from dad, L from mom.
   - L from dad, S from mom.
   - S from dad, L from mom.
   - S from dad, S from mom.

   Only in the last case in which you get the short eyelash gene from both parents will you have short eyelashes. 25%.

3. Let's first translate that question. If you see someone with short eyelashes, can you tell what their genotype is?

   If they have short eyelashes, then neither of their genes can be L. Hence, their genotype must be SS.

4. If I square zero, I get zero as an answer. If I square one, I get one as an answer.

5. Zero still works. The other possibility is three.

6. The only number that works is twenty-two.

The Bridge
from Chapters 1 to Chapter 22

first try

> Goal: Get 9 or more right and you cross the bridge.

1. Darlene wanted to cook a nice dinner for Joe. She knew if she asked him what he would like for dinner, he might ask for a baloney-and-ketchup sandwich and a big bottle of Sluice to drink. (That was also Joe's favorite breakfast.)
    Twenty of the 500 pages in the *Future Brides* magazine were devoted to "Meals You Can Cook for Him." What percentage is that?
2. Thirty percent of the 500 pages contained articles on dieting. How many pages were devoted to dieting?
3. One item that she knew Joe would like was steak with butter sauce. Before she cooked the steak, it weighed 2 lbs. and 7 oz. When she cooked it, it lost 11 oz. How much did the cooked steak weigh? (16 oz. = 1 lb.)
4. What number makes $x + 17 = 23$ true?
5. Suppose the dad had the genotype LS (where L is the gene for long eyelashes and L is dominant over S), and the mom had the genotype SS. What percentage of their children would be expected to have long eyelashes?
6. If a pair of fake eyelashes cost 59¢, how much would 7 pairs cost? (Express your answer using a dollar sign.)
7. Change $8\frac{1}{7}$ into an improper fraction.
8. What is the cardinal number associated with {2, 7, 8}?
9. Steak with butter sauce is very easy to make. You cook the steak and put it on a platter. Then you melt a cube of butter and pour it over the steak. What is the volume of this cube of butter? (The measurements are in inches.)
10. True or False: $3\pi > 9$

The Bridge
from Chapters 1 to Chapter 22

second try

---

1. For dessert, Darlene gave Joe a giant ice cream cone. It was seven feet tall and had a radius of three feet. (This was to show Joe how much she loved him.) She filled it up to the top with the richest ice cream she could find. (Darlene didn't fill it over the top since it would probably melt by the time Joe finished the steak.) What was the volume of the ice cream she used? (Use 3 for π.)
2. Darlene wanted to buy 70 cubic feet of ice cream. This is more than she needed to fill the cone, but she wanted a little extra in case she spilled some. How many gallons of ice cream did she buy? (1 cubic foot is approximately 7.5 gallons.)
3. Joe finished the whole steak in 3 minutes and 14 seconds. (He didn't leave any for Darlene.) She had thought that such a big steak would have taken him 10 minutes to eat. How much less time did Joe use than Darlene's estimate?
4. Out of the twelve people that Darlene knew, only Joe could have finished the steak that quickly. What percent is that? (Round your answer to the nearest percent.)
5. Retina are located in what part of the body?
6. What is one-fifth as a percent?
7. Is {Joe} a subset of the set of all the people that Darlene knows?
8. $(4¼)^2 = ?$
9. After Joe finished his giant steak, he asked for a straw so that he could drink up the butter that was still on his plate. (He knew it was impolite to just hold the plate up to his lips and drink it.) Darlene didn't have any straws in the house and suggested he use his spoon. "No use wasting money," he said. How much money did he "save"? (He drank five ounces. Each ounce cost $0.42.)
10. $7^3 = ?$

The Bridge
from Chapters 1 to Chapter 22

third try

---

1. "What's for dessert?" Joe almost shouted. Darlene had hardly had time to eat any of the parsley she was working on. She had bought a cup of parsley and had only been able to consume 2 ounces of it before Joe had finished his steak. What percentage of her dinner had she eaten? (1 cup = 8 ounces)

2. Darlene dropped her parsley and ran into the other room to get Joe's 7 foot tall ice cream cone. When she tried to wheel it into the room, she realized that the doorway was only 6' 8" tall. How much taller was the cone than the doorway?

3. She called to Joe, "Honey, I've got a real surprise dessert for you. Come in here and feast your eyes." Joe got up and in seven seconds he was standing next to the ice cream cone. He walked at the rate of 1.8 feet per second. How far did he walk?

4. Darlene had had $120 in her checking account and had spent 36% of it on this dinner for Joe. How much had she spent?

5. $\frac{5}{6} - \frac{4}{5} = ?$

6. Is Joe an autotroph? Explain your answer. Don't just say yes or no.

7. Joe had a 3-pound brain in a 150-pound body. What percent of his body was his brain?

8. What is one-tenth as a percent?

9. Name a subset of {5, 8}.

10. $\frac{5}{6} \times \frac{4}{5} = ?$ (Reduce your answer as much as possible.)

The Bridge
from Chapters 1 to Chapter 22

fourth try

1. Joe can think of 3897 foods he would rather eat than parsley. Is 3897 an ordinal number?

2. Darlene doesn't especially like parsley, but she eats a lot of it to stay slim. If it takes her 7.1 seconds to chop up an ounce of parsley, how long would it take her to chop up a pound of it? (1 pound = 16 ounces)

3. If Joe weighed 150 pounds and 5 ounces before he began his meal, how much would he have to eat in order to weigh 160 pounds? (1 pound = 16 ounces.)

4. When Joe looked at that giant ice cream cone, he turned to Darlene and asked, "Do you have an ax? I think I'll need to chop it down a bit before I eat it."

   She said, "I have two of them. I was hoping to save them to use on our wedding cake,* but you can use them now. I have an eleven-ounce ax and one that weighs three-fourths of a pound. Which one would you like?"

   Joe asked for the heavier one. Showing your reasoning, determine which one is the heavier ax.

5. Joe cut a hole in the side of the ice cream cone and watched the ice cream slowly ooze out. Darlene ran to the store to get a bucket so that the ice cream wouldn't get all over her carpet. She ran at 8 feet per second for 78 seconds. How far away was the store?

6. $\frac{47}{91} + \frac{8}{91} = ?$

7. What is the common name for pinna?

8. $(3\frac{1}{3})^2 = ?$

9. Buckets are $3.60 each. The sales tax is 7%. How much is the tax?

10. $(\frac{1}{2})^5 = ?$

---

* Darlene was picturing a very large wedding cake—the kind you couldn't just cut with a knife. His-and-Hers silver axes work perfectly.

The Bridge
from Chapters 1 to Chapter 22

fifth try

1. When Darlene got back to her apartment, she found Joe sitting on the floor. He announced, "I think I'm full." In all the years that Darlene had known Joe, this is the first time she had ever heard him say that.

She kidded him, "Would you like the rest to take home in a doggie bag?" That comment put Joe "over the top." He reached for the bucket and retched.

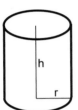

The bucket was in the shape of a cylinder and had a radius of one foot and a height of two feet. Find its volume. (The formula for the **volume of a cylinder** is simpler than the volume of a cone. It is just $V = \pi r^2 h$.) Use 3.1 for π.

2. There are two meanings of the word *retch*. The secondary meaning is to throw up. The primary meaning is to have the spasm in your stomach but not bring anything up. Joe went with the primary meaning. His face was red when Darlene brought the bucket. Then it turned white. When it turned blue, Darlene didn't think he was just being patriotic. She called 9-1-1. The ambulance took 6 minutes to get to Darlene's. Express 6 minutes in hours. (Use a conversion factor.)

3. The ambulance went at 50 miles per hour. How far did it go?

4. By the time the ambulance got there, Joe had switched to the secondary meaning of *retch* and was feeling much better. He had filled three-fourths of the bucket. (See question one above.) What volume had Joe lost?

5. Which of these three is the largest? 1^{35}, $0^{3979622}$, or 2^3

6. What is one-eighth as a percent?

7. The ambulance normally cost $200 per trip. Darlene got an 18% discount because her apartment was on the first floor. What was the price after the discount?

8. $\frac{2}{7} + \frac{1}{8}$ 9. $\frac{2}{7} - \frac{1}{8}$ 10. $(2\frac{4}{5})^2$

118

Chapter Twenty-three
Blood

Fred was walking back to the Math building where his office was. He was thinking about solving equations like $35x^6 - \pi x^5 + 4.6x^4 = 0$, and he wasn't really paying attention to what was going on around him. He had come to KITTENS University when he was nine months old, and he was now 5½ years old. So in the 4¾ years* he had been here, he had learned his way around pretty well.

In the campus park, there is a path called Harvey's Circle.

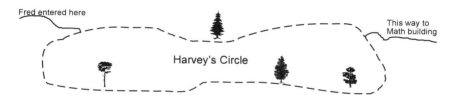

Lost in thought, Fred just kept going around Harvey's Circle path. Anyone who was watching must have wondered. This little kid, carrying a plant that was a foot taller than he was, was going round and round this path mumbling to himself $35x^6 - \pi x^5 + 4.6x^4 = 0$. He passed those four trees in the park many times.

Years ago, the biology students at KITTENS University constructed that path. They painted the path red and dedicated it to William Harvey.

While Fred is going around in circles, let's talk about William Harvey for a moment.

∗ How did I get the 4¾ years? You can't just subtract the 9 months from the 5½ years since they are in different units. First, I changed 9 months into years. Using a conversion factor: $\frac{9 \text{ months}}{1} \times \frac{1 \text{ year}}{12 \text{ months}} = \frac{9}{12}$ years $= \frac{3}{4}$ years.

Then I could subtract: $5\frac{1}{2} \quad 5\frac{2}{4} \quad 4\frac{4}{4} + \frac{2}{4} \quad 4\frac{6}{4}$
$\qquad\qquad\qquad\qquad\quad -\frac{3}{4} \quad\; -\frac{3}{4} \quad\quad -\frac{3}{4} \qquad -\frac{3}{4}$
$\qquad\qquad\qquad\qquad\qquad\qquad\qquad\qquad\qquad\qquad\qquad\;\; 4\frac{3}{4}$

Chapter Twenty-three   Blood

Harvey was English.

He graduated from a university in Italy.

In 1628 he wrote a book in Latin. (Latin is not a country.)

The book was published in Germany.

The book talked about how blood travels around the body in a circle. (That's why the biology students made the path in a circle and painted it red.)

Before Harvey, almost everyone believed . . .

Wait! Stop! I, your reader, am starting to detect a pattern.

First you had Jan Baptist van Helmont doing his science project with dirt and a willow tree overturning centuries of scientific thought.

Then you had Gregor Mendel doing his pea planting overturning centuries of scientific thought.

Now you got Bill Harvey . . .

Bill?

Okay, William Harvey.

Yes. For centuries before 1628, almost everyone believed what a Greek named Galen said about how blood moved in a body. Galen, who died in A.D. 201, thought that blood was made in the liver, then went through the heart, then through the arteries, and into the tissues (the cells) where it was absorbed! For 1400 years this was "common knowledge." Everyone knew it because Galen said it.

Scientists in the 1600s had been poking around inside of dead bodies (cadavers) and could estimate the volume of blood that the heart pumped with every beat. They knew how fast a living heart beat.

Say your heart pumps out 2 ounces of blood with every beat. Suppose your pulse is 72 beats per minute. Using conversion factors . . .

$$\frac{2 \text{ ounces}}{\text{beat}} \times \frac{72 \text{ beats}}{\text{minute}} = \frac{144 \text{ ounces}}{\text{minute}}$$

$$\frac{144 \text{ ounces}}{\text{minute}} \times \frac{60 \text{ minutes}}{\text{hour}} = \frac{8640 \text{ ounces}}{\text{hour}}$$

$$\frac{8640 \text{ ounces}}{\text{hour}} \times \frac{1 \text{ cup}}{8 \text{ ounces}} = \frac{1080 \text{ cups}}{\text{hour}}$$

Chapter Twenty-three Blood

$$\frac{1080 \text{ cups}}{\text{hour}} \times \frac{1 \text{ quart}}{4 \text{ cups}} = 270 \text{ quarts per hour.}$$

Harvey couldn't believe that all this blood could be absorbed by the tissues at that rate. It was too fast. Everyone around him said, "Who are you to contradict the great Galen?"

They knew about arteries and veins. Arteries were the big "pipes" that carried bright red blood, and veins were the smaller "pipes" that carried blood that was dark red. In cadavers they could see that both arteries and veins were attached to the heart. Galen said that blood only passes through the heart once, and that these arteries and veins carried blood away from the heart.

"Hey," thought Harvey. "This is easy to check." He found that severed (cut) arteries bleed only from the side of the cut that is toward the heart. Severed veins only bleed from the side of the cut that is away from the heart.*

Of course, all the Galen supporters said to Harvey, "Ha! Ha! Ha! You're telling us that 270 quarts per hour of blood flow out from the heart through the arteries, and then somehow all this blood is magically transferred over to the veins and flows back to the heart?"

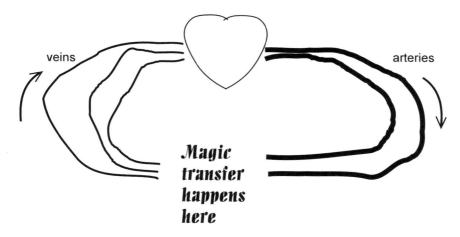

"I don't know how to explain the transfer," said Harvey, "but it's easy to show that arteries carry blood away from the heart, and veins carry it back toward the heart."

---

* Of course, this experiment wasn't done on cadavers. Please don't ask for details.

Chapter Twenty-three — Blood

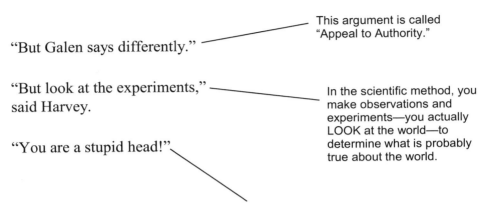

Poor old William Harvey. He had the overall picture right. He just couldn't explain how the blood got transferred over from the arteries to the veins so that it could make its trip back to the heart. That must have been frustrating for him.

Then, four years after he died, along came Marcello Malpighi* with a microscope. Microscopes were really new then, and you couldn't even purchase one on the Internet yet. (This was in 1661.)

Using his microscope, Malpighi found the missing link between the arteries and the veins. These tiny little "pipes" were so small that if you took a piece of muscle the size of a pinhead (•), there might be 750 of

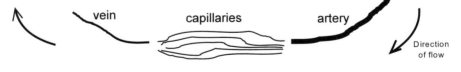

---

* You can imagine what ad hominem abusive remarks Marcello Malpighi received when he was in elementary school. First of all, they might have made fun of his initials, MM, which are the same as Mickey Mouse. Secondly, they might have called him "bad piggy." The prefix *mal* means "bad." Think of *malfunction* = to function poorly. Or *maladjusted* = not well adjusted.

Chapter Twenty-three Blood

them passing through that dot.* Since MM was Italian, he might have used the Italian word for hair, which is *capelli*, to describe these hairlike vessels. But, luckily, (since not all of us speak Italian), he used the Latin word for "hairlike" which is **capillaries**.

The combined diameters of the capillaries are about ten times the diameter of the artery that goes to those capillaries. That means that blood, which was rushing through the arteries, s-l-o-w-s way down. This is important because it is the capillaries which deliver the good stuff (the food and oxygen) to the cells. With the blood flowing leisurely by the cells, it gives the cells a chance to pick out the goodies they like.

Blood goes in a circle: heart → arteries → capillaries → veins → heart. That's why they call it blood circulation. (circulation = moving in a circle)

In the many centuries before 1600, almost everyone thought plants ate dirt.

In the many centuries before 1900, almost everyone thought that the traits of parents were blended in their kids.

In the many centuries before 1661, almost everyone thought blood took a one-way trip from heart to the tissues (cells).

Your Turn to Play

1. The smaller the body, the faster the heart rate tends to be. Kids have a pulse around 80 beats/minute. Infants clock in around 120 beats/minute.
   The heart of an elephant weighs around 50 pounds and beats around 30 times/minute. Using conversion factors, compute how many times an elephant heart beats in a year.
2. A canary's heart rate is about 1,000 beats/minute. Approximately how many beats per second is that?
3. If an elephant's heart weighs 50 pounds and a canary's heart weighs 3 ounces, how much more does an elephant's heart weigh than a canary's? (1 lb. = 16 oz.)

∗ Joe, whom you've met in the Bridges, once said to Darlene, "Those little pipes must be microscopic."

4. Negative numbers are numbers that are less than zero. For example, $-6 < 0$

When kids in school first learn about subtraction, they learn $8 - 5 = 3$. When they get bored, they ask their teacher, "What is 5 takeaway 8?"

If the teacher wants to silence the student, the teacher will say something like, "If you have 5 moose it doesn't make any sense to takeaway 8 moose."*

But if the teacher wants the kids to be curious and to learn, the teacher will point to cases where negative numbers make sense: "If it is 5° outside and the temperature drops by 8°, what is the new temperature?"

The answer is $-3°$.

What does $100 - 150$ equal?

5. $6 - 9 + 5 + 7 - 20 = ?$

······COMPLETE SOLUTIONS······

1. $\dfrac{30 \text{ beats}}{1 \text{ minute}} \times \dfrac{60 \text{ minutes}}{1 \text{ hour}} \times \dfrac{24 \text{ hours}}{1 \text{ day}} \times \dfrac{365 \text{ days}}{1 \text{ year}} = 15{,}768{,}000 \text{ beats/year}$

2. $\dfrac{1000 \text{ beats}}{1 \text{ minute}} \times \dfrac{1 \text{ minute}}{60 \text{ seconds}} = 60\overline{)1000} = 16\dfrac{2}{3} \doteq 17 \text{ beats/sec}$

$\begin{array}{r}1640/60\\\underline{-60}\\400\\\underline{-360}\\40\end{array}$

$\doteq$ means "which rounds off to"

3. $\begin{array}{r}50 \text{ lbs.}\\-3 \text{ oz.}\\\hline\end{array}$ $\begin{array}{r}49 \text{ lbs.}16 \text{ oz.}\\-3 \text{ oz.}\\\hline 49 \text{ lbs.}13 \text{ oz.}\end{array}$

4. $100 - 150 = -50$.

5. $6 - 9 + 5 + 7 - 20 = -11$

---

* This was the same teacher, who when asked, "What is one-half of five?" would respond, "If you have 5 kisses, you can't take half of them."

Chapter Twenty-four
Staying Alive

As Fred walked around Harvey's Circle thinking about solving the equation $35x^6 - \pi x^5 + 4.6x^4 = 0$, he suddenly realized that he hadn't done his morning jogging today. He started running. And kept thinking about $35x^6 - \pi x^5 + 4.6x^4 = 0$.

Biology is the study of living things. Biology books love to devote chapters to cells, to digestion, to photosynthesis, to the five kingdoms,* to genetics, etc. But virtually none of them devotes a chapter to staying alive. It seems kind of silly to omit that topic in a biology book.

I, your reader, have a small question. How can you include a chapter on staying alive when everyone else doesn't?

Easy. This biology book isn't like all the others.

Okay. I have another question. I think you made an English error. And since you say this biology book "isn't like all the others," I am free to ask. Seven lines ago you wrote, ". . . virtually none of them devotes a chapter. . . ." Shouldn't it be "devote," not "devotes"?

If you just look at the "them devotes" part of the sentence, it does sound funny. In English, the subject of the sentence and the verb have to agree. The "of them" is a prepositional phrase. It is not the subject of the sentence. The subject is the word *none*. The word *none* is singular.

Let's build it up slowly . . .

step 1: <u>Joe</u> <u>devotes</u> a chapter of his autobiography to eating.

step 2: <u>Joe</u> of all writers of autobiographies <u>devotes</u> a chapter to eating.

step 3: Of all the regular biology books, <u>none</u> <u>devotes</u> a chapter to staying alive.

step 4: <u>None</u> of them <u>devotes</u> a chapter to staying alive.

As Fred jogged past the four trees on Harvey's Circle, his goal was not just to stay alive, but to stay fully alive. In a word, to stay *healthy*.

---

* (1) Monera—single-celled bacteria. (2) Protista—single-celled or multi-celled with much more internal complexity than Monera. (3) Fungi–multi-celled and feed by extracellular digestion. (4) Plantae–plants. (5) Animalia—animals.

Chapter Twenty-four   Staying Alive

Everyone* knows that if you are in a coma, you can't think much about solving $35x^6 - \pi x^5 + 4.6x^4 = 0$, or about pizza, or about where you are.

I thought a coma was a piece of punctuation that looked like → ,

No that's a comma. Coma = being unconscious for a long period of time and unable to feel anything.

Fred finally took the exit off of Harvey's Circle and headed toward the Math building. He ran up the stairs to the third floor where his office was. The hallway on the third floor was lined with vending machines.

Fred was now five and a half years old. He had been without parental supervision for five years. (Full details of his early life are in *Life of Fred: Calculus*.)

In those five years no one had told him about the facts of life concerning foods.

Fred thought that the stuff in the vending machines were like things on a restaurant menu. They all fit into three categories:

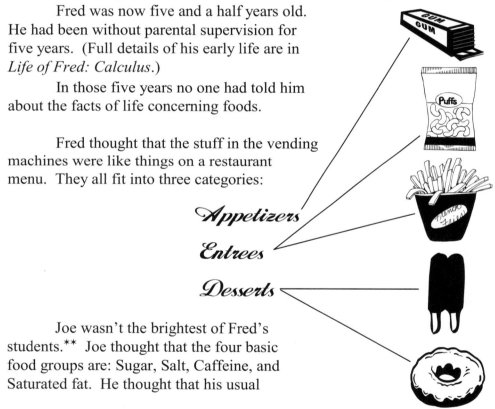

Appetizers

Entrees

Desserts

Joe wasn't the brightest of Fred's students.** Joe thought that the four basic food groups are: Sugar, Salt, Caffeine, and Saturated fat. He thought that his usual

---

★ except those in a coma [CO-muh where CO rhymes with DOUGH]

★★ An example of litotes. (LIE-teh-tease) An understatement. Litotes is often expressed as "the negative of the opposite." For example, instead of saying that our national debt is huge, you might use litotes and say that our national debt isn't exactly small. If I hadn't used litotes, I would have described Joe as one dumb bunny.

Chapter Twenty-four Staying Alive

lunch of a greasy triple burger and a quart of Sluice soft drink included all the four basic food groups. The hamburger had the salt and saturated fat. Sluice contained the sugar and the caffeine.*

Joe wasn't what you would call a candidate for living a long life. (litotes)

Excess salt tends to drive up blood pressure. High blood pressure is not something you can usually feel. No painful symptoms. The technical name for high blood pressure is **hypertension**. Besides eating excess salt, you can increase your chances for hypertension by being overweight, drinking alcohol, and having a life-style filled with stress, deadlines, and crises.

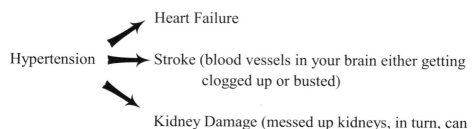

Salt   *Stout*   *Spirits*   *Stress*
        (*or Stuffing yourself*)

I, your reader, have one little question. If hypertension doesn't usually have any painful symptoms, why should I care?

Good point. I didn't get finished drawing the arrows. So far we have: (Salt, Stout, Spirits, Stress) ➝ Hypertension. From there we go to:

Hypertension ➝
- Heart Failure
- Stroke (blood vessels in your brain either getting clogged up or busted)
- Kidney Damage (messed up kidneys, in turn, can lead to even higher blood pressure)

---

* Words like *caffeine*, *science*, and *society* don't seem to follow the *i before e, except after c* rule. And *Einstein* violates the rule twice!

Chapter Twenty-four   Staying Alive

In some stores that have pharmacies, they sometimes have machines to check your blood pressure. If you are old enough to use them, you can sit down and stick your arm in the cuff, and the machine will tell you your blood pressure.

Your Turn to Play

1. Actually, the machine gives you two numbers—the high and the low numbers. That's because when your heart beats the pressure goes up. When it "rests" between beats it goes down. The machine has to figure out both numbers.

The two numbers are often written with a slash between them. For example, 125/82. This is not a fraction.

Let's look at the higher number for a moment. The official name for that number is the **systolic** reading. (sis-STOL-ick) *Systolic* is almost the last word in the dictionary under the *S* category. The city in Hungary called Szekesfehervar comes after *systolic* but it comes up much less frequently than *systolic* in everyday conversation.

Many books say that the super best, most wonderful, optimal systolic readings are those that are "<120." Your question: What does "<" mean?

2. If your systolic reading is typically >140, you are entering the hypertension zone. For fun, give an instance of when your reading was over 140 in which you did not have hypertension.

3. 120 is what percent of 140?

. . . . COMPLETE SOLUTIONS . . . .

1. The biologists have borrowed our "less than" sign from mathematics. Written out, it would be, "Optimal systolic readings are less than 120."

2. Your answer may be different than mine. There are many things that can send your blood pressure soaring that have nothing to do with the disease of hypertension. I picture myself sitting at that blood pressure machine, and over the loudspeakers they announce that they are giving away one hundred free pizzas at the other end of the store. My body would be gearing up to do some running.

3. 120 = ?% of 140. You divide the number closest to the *of* into the other number. $140 \overline{)120.000}^{.857}$ $0.857 = 85.7\% \doteq 86\%$

Chapter Twenty-five
Solving Algebra Equations

Since Fred still had a piece of french fry in his pocket, he didn't stop at the vending machines in the hallway. He could make several meals out of that one fry.

He looked down the hallway. His office door was open. All the dirt in his office was gone. There was a note on his desk:

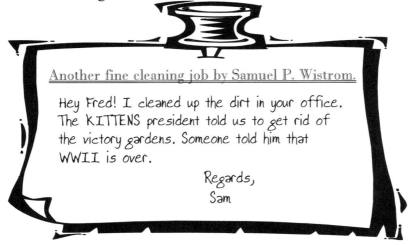

Another fine cleaning job by Samuel P. Wistrom.

Hey Fred! I cleaned up the dirt in your office. The KITTENS president told us to get rid of the victory gardens. Someone told him that WWII is over.

Regards,
Sam

Not a trace of dirt was left. The books on the shelves on all four walls were in perfect order. My sleeping bag is gone! Fred thought. He headed down to the basement and knocked on the door marked: Samuel P. Wistrom, Educational Facility Math Department Building KITTENS University, Chief Inspector/Planner/Remediator for offices 225–324. (His old sign had read Sam the janitor.)

"Hi Sam," Fred began. "I saw your note. You sure cleaned out my office as clean as a whistle."

"Yup. As clean as a whistle," Sam answered.

"I have a question. Do you know where my sleeping bag is?"

"I ain't seen it. But my new vacuum cleaner sure do work wonders, don't it," Sam answered.

"You vacuumed the dirt?" Fred couldn't believe what he was hearing. "That doesn't seem possible. Wouldn't your vacuum cleaner fill up? There was about 112 cubic feet of dirt in my office."

Sam scratched his head. "I wouldn't know about any feet in your office, but my new vacuum cleaner, she could handle them if she found them."

Sam opened his closet door and showed Fred his new machine. It was a giant hose that was two feet in diameter. Sam explained how he stuck one end in Fred's office and ran the hose down two flights of stairs and into the dumpster. After turning on the switch, the room was clean, and the dumpster was full.

Fred now knew where his sleeping bag was.

Fred looked at the cuckoo clock on Sam's wall. *Oh my! It's almost time for my first class.* Fred raced up the stairs to his office and put on his bow tie. He liked to wear a bow tie when he was teaching. He thought it made him look older. (That's an important consideration when you are teaching at a university and are only five and a half years old.)

As he raced across the campus quad,* he was thinking about what he was going to teach his eight o'clock beginning algebra class. Earlier in the semester, the class had learned about sets and about negative numbers. It was time for them to learn about solving equations.

---

* *Quad* has several meanings. If people say that they are a quad, that could mean that they were one of four born in a single pregnancy or that they were quadriplegic. What Fred was racing across was a quadrangle on a college campus. That's a rectangular open space bordered by buildings. Often quads have lots of grass and trees so students can have a comfortable place to lie down and read math.

Okay. To be fair, some students on a quad are not reading math. Those students who are not reading math are reading *Divina Commedia*: *Nel mezzo del cammin di nostra vita.* . . . [Midway upon the journey of life. . . .]

Chapter Twenty-five   Solving Algebra Equations

Some algebra instructors would start out with "baby" equations like x + 5 = 8, and would ask the class to find a number that makes that true. Everyone in the class (except maybe Joe and Darlene) would say, "Duh. That's so easy. The answer is 3." And the students would think that algebra is really worthless.

As Fred raced across the quad, he was composing his lecture in his head. He imagined that the first equation he would give them would be:

$$4x + 16 + 5x = 18 + 3x + 19 + 21$$

Nobody in the class would say, "That's too easy." From the very beginning, you would really *need* algebra to find the value of x that makes that equation true.

A real equation needing real algebra.

First, look at the left side of the equation: 4x + 16 + 5x
Combine together the 4x and the 5x: 9x + 16

Next, look at the right side of the equation: 18 + 3x + 19 + 21
Combine together the 18, 19, and 21.   58 + 3x

So 4x + 16 + 5x = 18 + 3x + 19 + 21 would become
       9x + 16   =   58 + 3x

The technical name for all this is **combining like terms**.

Your Turn to Play

Please take out a piece of paper and write out the answer to each question before you go and look at the answer. The object of the game is not just to turn pages in this book, but to *learn* stuff.
1. Combine like terms in the equation: 7x + 20 + 9x = 47 + 5x + 61
2. Combine like terms in the equation: 83 + 3x + 7 = 16 + 66x + 27
3. Combine like terms in the equation: 234 + x + 976 = 7x + 12

Chapter Twenty-five Solving Algebra Equations

········**COMPLETE SOLUTIONS**········

1. 7x + 20 + 9x = 47 + 5x + 61 becomes
   16x + 20 = 108 + 5x
2. 83 + 3x + 7 = 16 + 66x + 27 becomes
   90 + 3x = 43 + 66x
3. 234 + x + 976 = 7x + 12 becomes
   1210 + x = 7x + 12   There is nothing you could do with
                           the right side of that equation.

Now that you can combine like terms, let me explain some of the English. When we talk about "like terms," we first need to know what a "term" is.

7 + 3x + 98696 has three terms. There are three things that are added together.

$6y^2 + \pi x$ has two terms.

1 + 2 + 4 + 8 has four terms.*

92374329074329 has one term.

69 − 2y has two terms. (Subtraction works the same as addition.)

$68632x^{397}y^{992111}z^{100}$ has one term.

✶ It's true that 1 + 2 + 4 + 8 has four terms. If you combined these like terms and got 15, then you would have changed an expression that had four terms into one that had only one term. Is that okay with you?
   Physicists tell us that the total amount of mass plus energy in the whole universe never changes. It's a constant. It is always the same.
   Mathematicians are a little less rigid. The number of terms in an expression can change if you combine together like terms.

Chapter Twenty-five   Solving Algebra Equations

The next part of the English is explaining what *like* means in "like terms."

Official definition: Two terms are like terms if they contain exactly the same variables (like x and y), with the same **exponents**.

These are exponents.

$6x^5y^9z^2$ and $980x^5y^9z^2$ are like terms.
$w^{907}x$ and $234 w^{907}x$ are like terms.

$77x$ and $77x^2$ are not like terms.
$6x^5y^9z^2$ and $8x^5y^8z^2$ are not like terms (since the exponents on the y are different).
15 and 4x are not like terms.

More English . . . instead of saying that y^5 is "y with an exponent of 5," you can say, "y to the fifth **power**."

Still more English . . .

x^2 is called x **squared**.
x^3 is called x **cubed**.
x^4 is called x to the fourth power.
x^{98} is called x to the ninety-eighth power.

Chapter Twenty-six
The Second Step in Solving Equations

As Fred headed across the quad thinking about presenting the equation $4x + 16 + 5x = 18 + 3x + 19 + 21$ to his beginning algebra class, he met a lot of students heading the other direction. Many of them greeted him with, "Good morning" or "Hi!" They all seemed to have extra bright smiles.

Then he realized that many of those students were his beginning algebra students. He ran to his classroom. It was empty except for Joe and Darlene, who were packing up their stuff and getting ready to leave.

Fred asked, "What's happening? Where are the students?"

Joe pointed to the blackboard and said, "Sam, the janitor, came in and wrote that."

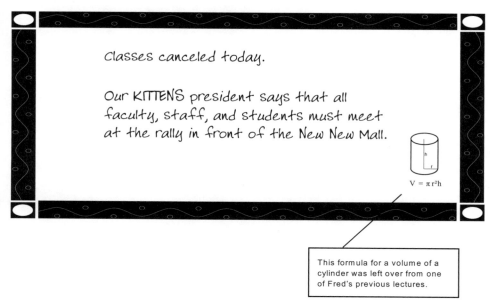

This formula for a volume of a cylinder was left over from one of Fred's previous lectures.

A zillion questions ran through Fred's mind:

1. What's the New New Mall? Did Sam just make a mistake in his writing?
2. A rally for what?
3. Why are they canceling classes? Education is supposed to be the first priority of any university.

Chapter Twenty-six   The Second Step in Solving Equations

4. Where is this New New Mall? I know where the New Mall is. I was just there this morning.
5. Why did the president cancel classes for the whole day?
6. Is it because the rally is going to last all day?

Hold it. I, your reader, also have a question. Is "zillion" really a word?

Yes. It was created in the early 1930s. It's in the dictionary.

What do you mean "created"? People don't just create words.

Where do you think our words come from? It's a scientific fact that ducks didn't create our words. Until the early 1930s, the word *zillion* didn't exist. Then somebody said, "zillion" to indicate some giant number like million, billion, or trillion.

photosynthesis

In contrast, there are things that existed before we ever thought of them. For example, if you take any right triangle*, it is always true that $a^2 + b^2 = c^2$. We didn't invent that.

(Neither did the ducks.)

a
c
b

> *Intermission*
>
> Why mathematics was the major I chose in college.
>
> If I wanted to study the stuff that ducks make (ducklings, noise, and duck poop), I could have chosen to be a biology major.
>
> If I wanted to study the stuff that legislators make (laws, regulations, and taxes), I could have studied law.
>
> If, instead, I wanted to study the things that are more permanent than poop and taxes, then mathematics was the perfect choice.

✶ A right triangle is a triangle with a right angle in it. A right angle is a 90° angle. A 90° angle is the angle at the corner of a square.

Fred said to Joe, "We are supposed to go to a rally at the New New Mall, but I don't know where that is. Do you know where it is?"

"Sure," said Joe. "Everybody knows. It's been on television. It is the same as the old New Mall, only newer."

Television Fred thought. The last time I watched television was when I was five days old.*

Fred, Joe, and Darlene left the classroom at the same time. Fred raced ahead at 7 feet per second. He didn't want to be late for the rally that the university president had announced. Joe and Darlene didn't care if they were late. They walked at 3 feet per second. How many seconds had elapsed when Fred was 800 feet ahead of Joe and Darlene?

This is called a "word problem" in algebra. By the magic we teach in algebra, this gets translated into the equation $7x = 3x + 800$. The hard part is doing the translating from words to equation. The easier part is solving the equation. Since this is just a pre-algebra book, we will just solve the equation.

Wait! Before you solve that equation, I need to know how you changed the word problem into the equation.

But that's like teaching your baby how to walk by entering your kid in a foot race.

I insist.

Or like eating a frozen pizza before you cook it.

This is my book. I bought it. Why aren't you going to do the harder part of algebra now?

Or like . . . okay. I'll put it in the next chapter. But only because you insist. I'm not sure I want to do it. It may be a little too hard at this stage of your math career.

Now you can get back to solving $7x = 3x + 800$.

Thank you.

The first step is to check if there are any like terms on either side of the equation that can be combined. The 3x and the 800 can't be combined—one has an "x" in it and the other doesn't.

∗ This happened in Chapter 6 of *Life of Fred: Calculus*, when Fred and his doll Kingie watched the STOCK MARKET CHANNEL. Watching television hasn't been a big part of Fred's life. (litotes)

Chapter Twenty-six The Second Step in Solving Equations

The 7x and the 3x can't be combined, because they are not on the same side of the equation—one is on the left side and one is on the right side.

So what do we do with 7x = 3x + 800 ?

We subtract 3x from both sides of the equation.
$$4x = 800$$

Hey. That's a neat trick. You had terms containing x on both sides of the equation, and now all of the x-terms are on one side.

The reason why we are allowed to do this is that *If you have two things that are equal, and you do the same thing to each of those two things, then the results will be equal.* That just makes sense. If, for example, you have two things that weigh* the same, and you double each of them, then the results will be equal. That rule, which is in *italics*, is so important that we will put it in a box:

> If two things are equal,
> and you do the same thing to each
> of those two things,
> then the results will be equal.

Wait. I have to ask. Why the accordions?

I like accordions. They remind me of my Uncle Charlie who used to play an accordion. Now please let me get back to the algebra.

Okay. I was just asking.

Thank you.
We started with   7x = 3x + 800.
We subtracted 3x from each side   4x = 800.
We divide both sides by 4   x = 200 and we're done.

---

* *Weigh* is another word that seems to violate the *i before e, except after c* rule.

Chapter Twenty-six   The Second Step in Solving Equations

Your Turn to Play

Special Rule for this *Your Turn to Play*: Solving equations is something you will be doing heaps and gobs of for a long time. No matter where you are educated in the world, being able to solve 7x = 3x + 800 is something that you will, in all probability, be expected to do. Your kids will expect that you know how to solve equations. To get into the Masters of Business Administration program at many colleges requires a knowledge of calculus—and calculus certainly requires algebra. The only place where a knowledge of solving equations *won't* be required is on the "exam" to get into Heaven. (A complete answer key to that exam can be obtained at no charge. Look in the yellow pages of your phone book between Chiropractors and Cigar Dealers.)

There are too many equations to solve in this *Your Turn to Play*. The special rule is that you work enough of these problems until you know how to do them while you are standing on your head and eating peanut butter. Learn the procedure well enough that you could teach it in a classroom to beginning algebra students. Then stop before it gets boring. Don't do all of them unless you want to.

1. 9x = 4x + 55
2. 3x = 287 + 2x + 178
3. 4x + 27 = 7x   (In this case, the x-term will end up on the right side.)
4. 13x + 19 + 2x = 21x + 1
5. x + x + 5x + 3 = 24
6. 32 + 6y + 2y = 24y (The variable doesn't always have to be x.)
7. The variable could be any letter of the alphabet. What would probably be the worst variable to use? (Your answer may differ from mine.)
8. 200 = 15x
9. 14 + 6w + 13w = 26w

   I'm getting bored. The equations below I haven't taught you how to solve. Good luck!

10. $x^2 = 16$
11. $y^3 = 27$
12. $x^x = 4$ (Side note: Even after you have had beginning algebra, advanced algebra, geometry, and trig, you won't be able to solve $x^x = 5$.)

······· COMPLETE SOLUTIONS ·······

1. $9x = 4x + 55$
Subtract 4x from both sides $5x = 55$
Divide both sides by 5 $x = 11$

2. $3x = 287 + 2x + 178$
Combine like terms $3x = 465 + 2x$
Subtract 2x from both sides $x = 465$

3. $4x + 27 = 7x$
Subtract 4x from both sides $27 = 3x$
Divide both sides by 3 $9 = x$

4. $13x + 19 + 2x = 21x + 1$
Combine like terms $15x + 19 = 21x + 1$
Subtract 15x from both sides $19 = 6x + 1$
Subtract 1 from both sides $18 = 6x$
Divide both sides by 6 $3 = x$

5. $x + x + 5x + 3 = 24$
Combine like terms $7x + 3 = 24$
Subtract 3 from both sides $7x = 21$
Divide both sides by 7 $x = 3$

6. $32 + 6y + 2y = 24y$
Combine like terms $32 + 8y = 24y$
Subtract 8y from both sides $32 = 16y$
Divide both sides by 16 $2 = y$

7. I personally think that "o"—or worse yet, capital "O"—is the worst. Is 5O fifty or "five-oh"? Some people hate "b" because it looks like a 6 when they write 6*b*. Others dislike "z" because it looks like a 2 when they write 2*z*. (When I taught, I always "crossed" my z's so they looked like *z*.)

8. $200 = 15x$
Divide both sides by 15 $13⅓ = x$

Chapter Twenty-six   The Second Step in Solving Equations

9. $14 + 6w + 13w = 26w$
Combine like terms $14 + 19w = 26w$
Subtract 19w and then divide both sides by 7 $2 = w$

10. Just "trying out numbers" you can see that $x = 4$ works for $x^2 = 16$.

11. Just "trying out numbers" you can see that $y = 3$ works for $y^3 = 27$.

12. Just "trying out numbers" you can see that $x = 2$ works for $x^x = 4$.

Chapter Twenty-seven
Words to Equations

Now I promised that I would try to explain how I went from the word problem to the equation five pages ago. I warned you that this explanation may be inappropriate at this stage in your math career. But you insisted.

Here goes.

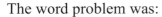

The word problem was:

    All three left the room at the same time.

    Fred was racing at 7 feet/sec.

    Joe and Darlene were walking at 3 feet/sec.

    How long before Fred got 800 feet ahead of them?

The equation I found was:
$$7x = 3x + 800.$$

Getting from the words to the equation is one of the more challenging parts of algebra, because there is no mechanical procedure. In the previous chapter, when I taught how to solve equations, I could break it down into steps: *First you combine like terms. Then you put the x-terms on one side of the equation and the numbers on the other side. Finally, if you have something like 6x = 30, you divide both sides by 6.*

In going from words to equations, all I can do is give lots of worked-out examples and some general suggestions.

 The most basic suggestion is that you read the problem enough times that you almost have the facts memorized.

 Draw some pictures or diagrams to help visualize what's going on.

 Let x equal the thing you are trying to find out. You actually write down on your binder paper: Let x = ...

Chapter Twenty-seven   Words to Equations

Let's put these suggestions into action. First, please read the word problem enough times that you can recite the facts of the problem. When you can do that, come back to this page and we'll draw a diagram.

Short Quiz to see if you really did reread the problem:
1. Did all three of them start out at the same time?
2. At the end of the problem, how far apart were they?
3. How fast was Joe going?
4. What is the problem asking for?
5. How fast was Fred going?

Okay. Now we draw a diagram.

The third ★ suggestion on the previous page was to let x equal the thing you are trying to find out. We are trying to find a *time*. We want to know how much time passed before Fred was 800 feet ahead of the couple.

On a piece of paper we write:
   Let x = the amount of time it took for Fred to get 800 feet ahead of J & D.

At this point you have done all three ★ suggestions. In algebra there are a dozen or so types of word problems. There are coin problems, consecutive number problems, d = rt problems, mixture problems, etc. For the student first learning algebra, there are lots of different kinds of problems. For teachers, the situation is different. They have seen hundreds of coin problems, and they all start to look alike.

A teacher will look at the word problem that we are working on and say, "It's a distance equals rate times time problem. d = rt."

Then the teacher will ask, "What do we know—what's given?"

You wrote on your paper: Let x = the amount of time it took for Fred to get 800 feet ahead of J & D. That's an important piece of information. That's why you wrote it down.

We know that Fred walked for x seconds.

Chapter Twenty-seven   Words to Equations

We also know that Fred walked at the rate of 7 feet/sec.

Bingo! We now know that Fred walked 7x feet. New information. Put it on the diagram.

How do we know he walked 7x feet? The easiest way is to use the d = rt formula. The rate (r) was 7 feet/sec, and the time (t) was x seconds.

We know that J & D also walked for x seconds.

We know that they walked at the rate of 3 feet/sec.

Bingo! We now know that they walked 3x feet. New information. Put it on the diagram.

Now look at the diagram you've drawn. Fred's distance from the classroom door is 7x. Fred's distance from the classroom door is also 800 + 3x.

So 7x = 800 + 3x.   Or, if you like, 7x = 3x + 800.

There. Now I've shown how to get from a word problem to its equation. I'm going to draw a line and get back to our pre-algebra and biology.

───────────────────────────────

Fred arrived at the rally ahead of Joe and Darlene.

I've drawn the line ━━━ . No more word problems for now. Soon enough you will be conquering beginning algebra, advanced algebra, geometry, trig, calculus. Soon enough you will be a junior at the university and select mathematics as your major. Soon enough you'll teach math at a place like KITTENS University for 45 years. Celebrate 50 years of marriage. Have grandkids, who will be reading their copies of *Life of Fred*. And then, soon enough, you will step into the Larger Life and leave these shadowlands.

Chapter Twenty-seven   Words to Equations

The old New Mall was in darkness

The New New Mall

The rally was held in front of the New New Mall. It looked a lot like the original New Mall, except that the lights were on. The university president was in the middle of giving his speech. He was on page 11 of a 47-page speech. Fred sat down in one of the metal folding chairs that Sam put out for the audience. Fred knew that he was supposed to be there since he was a member of the faculty, and the president had canceled classes today so that everyone could attend this rally.

Your Turn to Play

1. The metal folding chairs were arranged by Sam in a rectangle. There were 47 rows with 18 chairs in each row. How many chairs were there? Please compute this without a calculator.

2. Fred looked around at the audience. There were six people there (including himself). What percent of the chairs were occupied? Round your answer to the nearest percent.

3. Everyone (faculty, staff, and students) was *supposed* to be at this rally. What percent hadn't shown up? Again, round your answer to the nearest percent.

4. $45 + 8x + 54 = 19x$

5. $17y - 45 = 2y$ (This equation is a little different from the ones you have had already. In the previous equations, you either *subtracted* the same thing from both sides or you *divided* both sides of the equation by the same thing. Now you will *add* the same thing to both sides.)

6. When students are first learning how to solve equations, you write out on the left side things like "Combine like terms" or "Divide both sides by 35." After a while, those comments are not really necessary. Just looking at the equations you can tell how you moved from one line to the next.

Chapter Twenty-seven — Words to Equations

YOU ARE NOW THE TEACHER.™ Supply the reasons for each line in the solution of $48 - 2z = 13 + 6z + 3$

?	$48 - 2z = 16 + 6z$
?	$32 - 2z = 6z$
?	$32 = 8z$
?	$4 = z$

·······COMPLETE SOLUTIONS·······

1. Forty-seven times eighteen is eight hundred, forty-six.

2. We are asked, "What percent of 846 is 6?"
?% of 846 = 6. We divide the number closest to the "of" into the other number.
$$846 \overline{)6.000} \quad 0.007$$
$0.007 = 0.7\%$, which rounds to 1%.

3. If 0.7% did show up, then $100\% - 0.7\% = 99.3\%$ didn't show up. $99.3\% \doteq 99\%$. ($\doteq$ means rounded off)

4.
	$45 + 8x + 54 = 19x$
Combine like terms	$99 + 8x = 19x$
Subtract 8x from both sides	$99 = 11x$
Divide both sides by 11	$9 = x$

5.
	$17y - 45 = 2y$
Add 45 to each side	$17y = 2y + 45$
Subtract 2y from each side	$15y = 45$
Divide both sides by 15	$y = 3$

6. The reasons are:
	$48 - 2z = 13 + 6z + 3$
Combine like terms	$48 - 2z = 16 + 6z$
Subtract 16 from both sides	$32 - 2z = 6z$
Add 2z to both sides	$32 = 8z$
Divide both sides by 8	$4 = z$

Chapter Twenty-eight
Breathing

The university president droned on. He was on page 18 of his oration. Fred looked around. Of the five other people in the audience, one was reading a book. One was sleeping. The other three were looking around at Fred.

The president continued, ". . . in the noble tradition of the independent spirit which pervades Kansas, and, no, not only Kansas, but all of these United States, we have gathered here to celebrate the flowering of achievement in the opening of this New New Mall, as it has so often been called. . . ."

Two in the audience were now asleep. The fellow who had been reading had shut his book, muttered, "too noisy," and left.

By page 29 in his speech, the two that had been asleep had woken up. They started a card game to pass the time.

". . . and as we gaze at this magnificent demonstration of human achievement—these eighty stores in this New New Mall—we stand in wonderment at. . . ."

Fred thought about the word *wonderment*. Then he thought of "hot air." Then of the president's lungs. The air that comes out of a person's lungs is hotter than the air that went in.

Fred had a zillion thoughts about breathing.

1. Breathing is so easy that I can do it in my sleep. Fred giggled a little when he thought of that. The two people playing cards looked at him momentarily and then went back to playing cards.

2. In Sunday School, I learned that Adam became a living being after he started breathing.

3. As the president breathes in, the air gets a Triple Treatment:

    i) It is warmed. The lungs like it that way.

    ii) It gets moistened on the way to the lungs.

    iii) It gets cleaned. The hairs in the nostrils filter out the bigger stuff, and the wet, slimy stuff (mucus) in my nose catches some of the germs, smoke and dust.

Chapter Twenty-eight   Breathing

4. Do older people have more hairs in their nose than kids?
5. Breathing air is like the tides. It goes in and it goes out.
6. If I were a fish, I'd have to get my oxygen out of the water instead of out of the air. Water is a lot thicker than air. It would be a lot of work to breathe in the water and then shove it out again. Fish use gills—the water goes in one way and out the other.

At this point, the university president was on page 36: ". . . as we gaze into the future—in a way not dissimilar to gazing into the past, except that it hasn't happened already—we note that the future is ahead of us."

Fred looked around. He was the only one in the audience now. Sam came by and started removing some of the 846 chairs.

Fred had heard all of the president's speeches over the years. What he started to notice is that most of his speeches had the same phrases in it, such as, "we note that the future is ahead of us."

Last year Fred heard, ". . . and as we gaze at this magnificent demonstration of human achievement—this new parking lot—we stand in wonderment at. . . ."

On the last page of his speech, "And now ladies and gentlemen, and without further ado, it gives me great pleasure to introduce the owner of this New New Mall. Let's give a big hand for Miss Sally."

Fred clapped. He didn't think a standing ovation* was appropriate at this point. He settled in his chair awaiting another speech. His essential life processes were approaching their **basal metabolic rate**. I.e.,** his heart rate, his breathing, his generation of body heat, his digestion, were all

---

* Ovation = long, loud applause. Not to be confused with *oration*, which is a formal public speech, which is often also long and loud.

** I.e. = *id est* (Latin) = that is (English)

Chapter Twenty-eight Breathing

near their minimums. Usually, people are at their basal metabolic rate when they first awake in the morning.

But Fred's metabolic rate was about to change radically. When Miss Sally, age 2, walked out onto the stage, she looked at Fred . . . and winked. She waved at Fred.

His heart rate shot up. His breathing became faster.

There are lots of things that can increase metabolic rate: exercise, digesting a meal, hard thinking, and mental shock.

Fred couldn't believe it. Little Sally not only owned the New New Mall. She had reopened it.

Sally looked at Fred and said, "Hello, Mr. Man. Do you wanna come and look at my mall?"*

Sally's mom got on the stage. She took Sally by the arm. "Where have you been? Jimmy and I have been looking all over for you." Sally didn't get a chance to tell her mother that she had been to her lawyer, her accountant, her bank, and the Federal Office Of Liberating Juveniles Operating Businesses (which is affectionately known as FOOL JOB).

The stage was empty. Sam asked Fred to get off his chair so that he could take it and put it with the other chairs.

Fred stood there. He was alone. The New New Mall lay in front of him. He wondered if Sally had made any changes.

The workmen were putting up a new sign at the entrance to the New New Mall:

∗ This was the shortest speech ever delivered at a KITTENS University event.

148

Chapter Twenty-eight   Breathing

Your Turn to Play

1. At her lawyer's, Sally had incorporated herself. The lawyer worked 1.7 hours at $270 per hour to consult with her and choose which state to incorporate in. (Different states have different laws. They chose a state with favorable tax rates.) How much did Sally owe her lawyer? Please do this without a calculator.

2. At her accountant's, Sally learned that she would have to be paying federal income taxes, state income taxes, property taxes, and taxes on her phone bill. She was told that if she hired any employees, she would have to pay taxes on their wages for some government program. She couldn't believe that and asked, "Aren't the people who receive the wages the ones who have to pay taxes on the wages?" The accountant laughed. You pay taxes on those wages, and they pay taxes on those wages. Sally wasn't laughing.

   The accountant's bill was $200. "Why is it so high?" she asked. He told her that his taxes on that $200 would be $80. He said, "I also have to pay federal income tax, state income tax, property tax, sales tax, gasoline taxes, a tax on my business inventory. . . ." What percent of that $200 bill went toward taxes?

3. Solve $55x + 18 = 19 + 164$

4. At the bank, Sally asked for a loan on the New New Mall. They said that there was already a big loan on the mall, and that they would be able to lend her 0% of the value of the property. If the mall was worth five million dollars, how much would the bank lend her?

. . . . COMPLETE SOLUTIONS . . . .

1. Working for 1.7 hours at $270 = 1.7 × 270 = $459.

2. Eighty is what percent of 200?   80 = ?% of 200. You divide the number closest to the *of* into the other number.

$$200 \overline{)80.00} \quad 0.40 = 40\%$$
$$0.40$$

3.
	$55x + 18 = 19 + 164$
Combine like terms	$55x + 18 = 183$
Subtract 18 from both sides	$55x = 165$
Divide both sides by 55	$x = 3$

4. 0% of anything is zero.

149

Chapter Twenty-nine
Still Breathing

The bank wouldn't lend Sally any money. They explained to her that if there weren't such a big loan on the mall already, they would be happy to make a loan. They told her that if the mall didn't have any liens on it, the only requirement would be that she would be able "to fog a mirror."

Sally said that she was "double confused."

First she asked, "Who is leaning on my mall?" They explained to her that the word was *lien*, not *lean*. A lien is the legal right for someone to take your property if you don't pay them.

At the age of two, Sally had never heard of the expression "to fog a mirror." That was her second point of confusion. She asked, but they were afraid to answer. They debated whether they would be sued if they taught the "facts of life" to a two-year-old. They told her that she should ask her mother about such things.

Sally frowned. She knew her mother was always too busy to answer her questions.

When Sally went to the Federal Office Of Liberating Juveniles Operating Businesses, they didn't care that the mall had so many liens on it. They didn't care that it didn't have a history of making money. Here was the paper they handed Sally:

FEDERAL GOVERNMENT
Application for Money

Name: _Sally_

Age, Weight & Height: _2, 30, 28_

Do you need lots of money? ☑ yes; ☐ no

Can you fog a mirror? ☐ yes; ☐ no

Chapter Twenty-nine   Still Breathing

When Sally turned in her paper, she asked the woman behind the counter what "fogging a mirror" meant.

"Honey, I don't know either. But everyone seems to check the 'yes' box." The clerk checked the box for Sally and put Sally's application for money on the top of a tall pile of applications.

---

a small essay
To Fog a Mirror

What you breathe in is different than what you breathe out. The air that you inhale is about 20% oxygen. The air you exhale contains about 16% oxygen.*

The air that you inhale is about 0.05% carbon dioxide. Exhaled air is about 4% carbon dioxide.

The amount of water vapor in the air depends a lot on where you live. If you live in a hot desert, there is little humidity. Women in those climates tend to use a lot more lip grease to keep their lips from cracking. If you live in the swamplands and it is raining, the humidity will be pretty high. There you get the three M's: moss, mold, and mildew.

So the amount of water vapor you breathe in varies considerably. But the air you breathe out is pure swampland. It is practically saturated. Put your nose up to a window and breathe out. You will make fog holes.

So in the old days, when they wanted to figure out whether you were dead, they would stick a mirror under your nose. If you could fog a mirror, that meant *you weren't dead*.

When Sally inhales, she fills her lungs with oxygen-rich air. Some of the oxygen is transferred to her blood. Then through the heart and exits through the largest artery in the body—the **aorta**. [a-OR-ta] Then on to the capillaries. Here the oxygen gets dropped off at the cells.

∗ Small math note . . . "In" is 20%. "Out" is 16%. That does *not* mean you "lose" 4% of the oxygen you breathe in. Four percent is one-fifth of 20%. You use up one-fifth of all the oxygen you inhale.

Chapter Twenty-nine   Still Breathing

Inside the cells the oxygen is combined with the 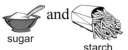 and

and [oil] —or, more accurately, the little sugars and amino acids that these have been broken down into by digestion. This chemical reaction gives off energy and carbon dioxide.

The carbon dioxide is picked up by the blood, transported through the veins, and is dumped into the lungs so that Sally can exhale it.

The blood traveling through the blood vessels reminds me of a truck which picks up bottles of the good stuff (oxygen) at the store (the lungs) and transports them to the factory (the cells) and then picks up the empty bottles (carbon dioxide) and takes them back to the store.

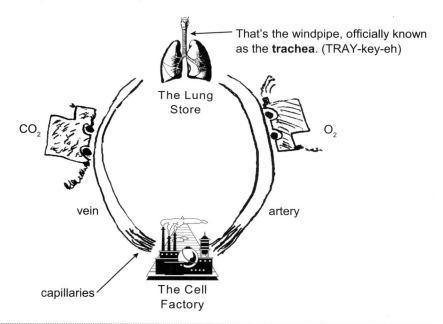

Your Turn to Play

1. The air that we breathe in is about 0.05% carbon dioxide (CO_2). The air that we exhale is about 4% CO_2. How many times more carbon dioxide is in our exhalation than our inhalation? (The **General Rule** is: *If you don't know whether to add, subtract, multiply or divide, first*

restate the problem with really simple numbers. If, for example, you breathed in 2% CO_2 and breathed out 10% CO_2, you could immediately say that the concentration had gone up by a factor of 5. Look to see how you got the 5—you divided—and then you know what to do in the original problem.

2. Solve $17 + 6y + 26 = y + 78$

3. On an average day, the average person takes 4500 grams of oxygen out of the air and puts it on the O_2 truck and sends it to the cells. Using the conversion factor, 1 gram = 0.0022 pounds, how many pounds of oxygen are transported each day?

4. $\frac{1}{8} \div \frac{3}{4} = ?$

......COMPLETE SOLUTIONS.......

1. 4% divided by 0.05% is the same as 0.04 divided by 0.0005.

    $0.0005 \overline{) 0.04}$

    becomes, after moving both decimals four places to the right,

    $5 \overline{) 400} = 80$ So the concentration increases by a factor of 80.

    English majors might say, "The concentration increases 80-fold."

2. $17 + 6y + 26 = y + 78$
    $43 + 6y = y + 78$ (Combine like terms)
    $6y = y + 35$ (Subtract 43 from both sides)
    $5y = 35$ (Subtract y from both sides)
    $y = 7$ (Divide both sides by 5)

3. $\frac{4500 \text{ grams}}{1} \times \frac{0.0022 \text{ pounds}}{1 \text{ gram}} = \frac{4500 \text{ grams}}{1} \times \frac{0.0022 \text{ pounds}}{1 \text{ gram}}$

    = 9.9 pounds each day. And one-fourth of that is used by the brain.

4. $\frac{1}{8} \div \frac{3}{4} = \frac{1}{8} \times \frac{4}{3} = \frac{4}{24}$ which reduces to $\frac{1}{6}$

The Bridge
from Chapters 1 to Chapter 29

first try

> Goal: Get 9 or more right and you cross the bridge and go on to Chapter 30.

1. Joe announced to Darlene that he was thinking of getting a dog. She responded with, "Why?" Seven-eighths of the time that Joe announced something, that was her response. What is seven-eighths as a percent?

2. Joe said that when he went fishing and caught a fish, he could have the dog jump into the water and fetch the fish. Darlene explained to him that he was confusing fishing with duck hunting where you shoot the duck and the dog goes and fetches the dead duck. Joe responded, "Oh," which was his response 88% of the time Darlene said anything to him. Change 88% to a fraction (reduced to lowest terms).

3. Joe said, "I want a dog to go along with us when we go fishing."

   "Why?"

   "So he could eat the food that I don't eat," said Joe.

   Darlene laughed to herself. She normally made six pounds of baloney-and-ketchup sandwiches for Joe. How many ounces is that? (1 pound = 16 ounces.)

4. Of those six pounds of sandwiches, Joe eats 98%, drops 1% overboard, and uses 0.5% for bait. How many ounces are left for the dog to eat?

5. Joe got out his bucket of pennies. It was 8" tall and had a radius of 5", and it was full. What was its volume? (Use 3 for π.)

6. Joe saw an ad in the newspaper: Our Special: Buy 6 dogs and spend $8 for leashes—for a total cost of $476. How much would a dog cost? (We let x equal the cost of one dog.)

   (Hint: That is the same as solving $6x + 8 = 476$.)

$V = \pi r^2 h$

7. How many terms does $2 + 4 + 6$ have?

8. Change $10\frac{1}{5}$ into an improper fraction.

9. What is the cardinal number associated with $\{1, 3, 5, 7\}$?

10. $7^3 \times 5^3 \times 0^8 = ?$

154

The Bridge
from Chapters 1 to Chapter 29

second try

---

1. This was the fifth time this week that Joe had thought of getting a dog. Is *fifth* a cardinal number?

2. Supply the reason for going from $3x + 17 = 5x + 9$
   to $17 = 2x + 9$

3. Combine like terms in the expression $6 + 7w + 8w + 9w^2$

4. True or False: $\pi - 3 > 0.1$

5. Joe's bucket of pennies had a volume of 600 cubic inches. If a cubic inch of pennies weighed two ounces, how much did his pennies weigh? Give your answer in pounds.

6. Joe wanted to buy a dog using his pennies. Using just the information given on this and the previous page, is it possible to determine how many pennies Joe has?

7. Joe announced to Darlene that he wanted to spend one-eighth of everything he owned to buy a dog. How much was he willing to spend?

Joe's total possessions:
    comic books $6.77
    fishing stuff $15.88
    clothes $19.98
    TV set $834.00
    cash $83.37

8. Reduce to lowest terms $\frac{24}{30}$

9. Change $\frac{24}{30}$ to a percent. (Note, this is the same fraction as the previous problem.)

10. $5\frac{2}{3} \div 2\frac{2}{3}$ (Simplify your answer as much as possible. Do not leave it as an improper fraction.)

The Bridge
from Chapters 1 to Chapter 29

third try

---

1. Suppose that a penny weighs eight grams.
Suppose Joe's bucket weighs 23 grams without any pennies in it.
Suppose that Joe's bucket and the pennies in it weigh 12,023 grams.
If we let x equal the number of pennies, then the equation will be
$$8x + 23 = 12{,}023$$
Solve this equation.

2. Joe spent one afternoon counting the number of pennies he had in his bucket. He counted 1492 pennies. (He does not count very accurately.) Express 1492¢ using a dollar sign ($).

3. Joe counts 4 pennies every 15 seconds. How many seconds would it take him to count 1492 pennies?

4. The answer to the previous problem, expressed in minutes, is 93.25 minutes. 93.25 minutes is equal to 93 minutes and how many seconds?

5. Joe and Darlene headed off to the pet store so that Joe could buy a dog. The first dog that they looked at was really cheap. But, neither Joe nor Darlene liked it very much. How much is this dog?

Sale! Normally $120 Now 80% off.

6. { 🐕, 🐕, 🐕 } is the set of dogs Darlene liked.

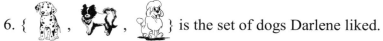

Joe preferred this set: { 🐕, 🐕, 🐕, 🐕 }. Are these two sets disjoint?

7. After an hour of thinking, Joe selected his favorite: 🐕 It was $36 plus a sales tax of 6%. How much is the sales tax on that dog?

8. Is $2^3 > 3^2$?

9. 🐕 had 17 square inches of skin and 8 fleas on every square inch. How many fleas did it have?

10. $4\frac{4}{5} \div 3\frac{4}{7}$

The Bridge
from Chapters 1 to Chapter 29

fourth try

---

1. When Joe brought 🐕 home, he decided it needed a little treat. He headed to the freezer and got out a carton of Sluice-Flavored™ ice cream. The height of the carton was 7" and the radius was 2". Using $\pi = 3.1$, find the volume of the carton.

2. "Nothing is too good for my doggie," Joe told Darlene as he emptied the entire carton of ice cream into a bowl. If the dog could finish three-fifths of the ice cream in 48 seconds, how long would it take to finish seven-eighths of the bowl? (Hint: You can compute this several ways. One way is to use a conversion factor. You know that $\frac{3}{5}$ bowl = 48 seconds. So this fraction $\frac{3/5 \text{ bowl}}{48 \text{ secs}}$ or this one $\frac{48 \text{ secs}}{3/5 \text{ bowl}}$ is equal to one.)

3. After the dog finished the whole bowl, it turned a little blue and didn't feel so good. 🐕 (If this is your book that you paid for with your own money, feel free to color the dog blue if you wish.) Joe picked up his blue dog and carried it out to the car. "We gotta get my doggie to the dog doctor," Joe said. He was walking at 6 feet per second for 18 seconds. How far did he carry his dog?

4. They stopped at a gas station and spent $17.35 for gas. He also spent seventy-two cents for a candy bar. How much did he spend?

5. Joe tried to feed the candy bar to the dog. He told Darlene that it would give it "quick energy." The dog wasn't interested. 🐕 Joe ate the candy bar (1/3 pound) plus 3/5 of a pound of "snack nibbles" that he had in the car. How much did Joe eat on the way to the vet's?

6. Eight is what percent of 32?

7. Nine percent of 99 is how much?

8. How much larger is $\frac{3}{4}$ than $\frac{2}{3}$?

9. Solve $32 + 3x = 5x$

10. $932 + 0.932 = ?$

(Recall the **General Rule**: *If you don't know whether to add, subtract, multiply or divide, first restate the problem with really simple numbers.*)

The Bridge
from Chapters 1 to Chapter 29

fifth try

1. Joe carried his dog 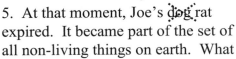 into the vet's office. "Can you fix my dog?" Joe asked. The vet had a deep look of concern. There were four reasons the vet looked worried.

First reason for worry: Can Joe pay for the veterinary work? The vet said to Joe, "I charge $3 per pound that the animal weighs plus a flat charge of $25. For example, when I treated a sheep last week, the bill was $226."

If we let x = the weight of that sheep, then the bill was 3x + 25. The equation is 3x + 25 = 226. Solve it to find the weight of that sheep.

2. The second reason the vet was worried: The animal looked nearly dead. The vet took its blood pressure. It was 16/5. What was the systolic reading?

3. The third reason for worry: Joe asked that his dog be *fixed*. The vet was incredulous that Joe would want a nearly dead animal fixed. In my dictionary, there are 19 meanings for the verb *to fix*. Joe was thinking of the first meaning of *fix*, which is to repair—as in "to fix a car." The vet was thinking of the 16th meaning: to neuter an animal. Is 16th an ordinal or a cardinal number?

4. The fourth reason the vet worried was that Joe had asked that his *dog* be fixed. The vet showed Joe a chart on the wall and explained to him, "This is the set of all dogs." Then he showed Joe a chart on another wall. "This is the set of all rats." Are these two sets disjoint?

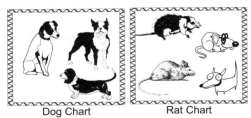

Dog Chart   Rat Chart

5. At that moment, Joe's dog rat expired. It became part of the set of all non-living things on earth. What is the union of the set of all living things on earth with the set of all non-living things on earth?

6. $(0.5)^2 = ?$ 7. Solve $12w - 31 = 4w + 9$ (Begin by adding 31 to both sides.)

8. $\pi \div \pi = ?$ 9. Change $33\frac{1}{3}$ into an improper fraction. 10. 6 is what percent of 60?

Chapter Thirty
How Oxygen Is Carried in the Blood

The truck that carries the oxygen from the lungs to the cells is a real truck, metaphorically speaking.* That truck is called **hemoglobin**. Oxygen doesn't dissolve very well in the fluid, watery part of blood. Without molecules of hemoglobin to truck the oxygen, your blood could carry about 60 grams of oxygen per day to your cells. In the previous *Your Turn to Play* we noted that your cells need about 4500 grams of oxygen per day.

Without the hemoglobin molecules, you are a dead duck. (metaphor)

The hemoglobin molecules aren't exactly small. (litotes) Their molecular weights vary from 17,000 to some that run into the millions. In contrast, the molecular weight of H_2O is 18. (Each hydrogen weighs 1, and oxygen weighs 16.)

---

* Language that uses words only in their literal meaning is *as dull as dishwater*. With the figurative use of words our communication *lights up*.

In the old days, everyone knew what the head of a person was. It had a nose, eyes, ears, a mouth, etc. Then someone invented golf clubs. Is it possible for a golf club to have a head? Not if every word must be taken in its literal meaning. What part of a golf club has a nose, eyes, ears and a mouth?

The literalist says, "the part of a nail that you hit with a hammer." Everyone else calls it the *head of a nail*.

Can air talk? Not literally. But *the heavens declare the glory of . . .* is a phrase heard often in synagogues and churches.

Even scientific papers make some use of figurative language. And even some math books—but finding one of them is *like finding a needle in a haystack*.

A **simile** (SIM-a-lee) is a comparison. It uses the words *as* or *like*. Three sentences ago, I wrote a simile. To find a good simile is *like pulling teeth*—it is a difficult thing to do.

A **metaphor** is more *hard-hitting* than a simile. A metaphor does not *pull its punches*. It simply states that something is something else. It doesn't *apologize* by using *as* or *like*. It simply *spits it out*.

The crowning moment in my life was when. . . .

Mice *stitched their tracks* across the snow.

Hemoglobin is *the truck* that carries the oxygen to the cells.

159

Chapter Thirty How Oxygen Is Carried in the Blood

In the middle of these huge hemoglobin molecules is the heme group. In the middle of the heme group is an iron atom. It's not a big part of the whole molecule. (litotes) The molecular weight of an iron atom is only 56.

But that little iron atom is the heart (metaphor) of that molecule. Oxygen from the lungs attaches to that iron atom and takes a ride to the cells. When it gets to the capillaries, the oxygen hops off the truck and enters the cells.

When oxygen is riding the hemoglobin truck, the molecule turns bright red. Cut an artery and you get bright red blood. After the oxygen hops off, the molecule becomes a darker purplish red.

A Bit of Wonderment

In the middle of the hemoglobin molecule is the heme group. In chemistry they drew it something like this.

Chemistry Notes:
  H = hydrogen
  N = nitrogen
  C = carbon
  Fe = iron
(They don't use I for iron, because I = iodine.)

At every vertex (corner) is a C. They don't write them all in so the diagram is easier to look at.

Now take a peek at the chlorophyll molecule. The big difference is that there is a magnesium (Mg) atom in the middle instead of an iron atom. Pop out the Mg atom and stick in a Fe atom and you are well on your way from the green stuff in plants used for photosynthesis to the red stuff in your blood that trucks oxygen from your lungs to your cells.

This is too weird.

160

Chapter Thirty — How Oxygen Is Carried in the Blood

We are not done with wonderment. Pop out the iron atom from the hemoglobin and stick in a copper (Cu) atom, and it turns from red to blue.

Instead of hemoglobin, we now basically have **hemocyanin**. Hemocyanin is the carrier of oxygen in arthropods.

The arthropod phylum:

That's why when you swat a fly, you don't get a big (red) bloody mess.

(You don't get a *big* mess, because flies don't carry a quart of hemocyanin. If they did . . . you would have to trade in your flyswatter for a baseball bat.)

Chlorophyll	green	Mg	photosynthesis in plants
Hemoglobin	red	Fe	oxygen transport in most animals
Hemocyanin	blue	Cu	oxygen transport in insects, spiders, crustaceans*

Once the oxygen hops off the hemoglobin molecule and enters the cell, the cell factory (metaphor) combines it with various food items that the blood has also delivered.

For example, here is a simple sugar called **glucose** ($C_6H_{12}O_6$). This molecule has six carbon atoms, twelve hydrogen atoms and six oxygen atoms.

The oxygen molecule, O_2, consists of two oxygen atoms.

We know that when we combine these two molecules we get carbon dioxide and water.

In chemistry the **skeleton equation** is

$$C_6H_{12}O_6 + O_2 \rightarrow CO_2 + H_2O$$

---

* Crustaceans = lobsters, crabs, shrimp

Chapter Thirty   How Oxygen Is Carried in the Blood

But $C_6H_{12}O_6 + O_2 \rightarrow CO_2 + H_2O$ is not possible. We have 8 atoms of oxygen on the left side and only 3 on the right side. The equation is not balanced. In the next chapter, we'll show you how to balance chemical equations. It's different than solving algebra equations.

Your Turn to Play

1. Solve $9x + 38 - 3x = 80$
2. $\left(\frac{4}{5}\right)^3$
3. $6\frac{2}{7} \div 2\frac{2}{3}$ (Remember to reduce your answer, if possible.)
4. Solve $4 + 4y = 1 + 7y$
5. Express $0.15 using the ¢ sign.
6. How many terms does $6x^2 + 398 + 7$ have?
7. How many subsets does {G, H} have?

. . . . COMPLETE SOLUTIONS . . . .

1.
$$9x + 38 - 3x = 80$$
$$6x + 38 = 80$$
$$6x = 42$$
$$x = 7$$

2. $\left(\frac{4}{5}\right)^3 = \frac{4}{5} \times \frac{4}{5} \times \frac{4}{5} = \frac{64}{125}$

3. $6\frac{2}{7} \div 2\frac{2}{3} = \frac{44}{7} \div \frac{8}{3} = \frac{44}{7} \times \frac{3}{8} = \frac{\overset{11}{\cancel{44}}}{7} \times \frac{3}{\underset{2}{\cancel{8}}} = \frac{33}{14} = 2\frac{5}{14}$

4.
$$4 + 4y = 1 + 7y$$
$$4 = 1 + 3y$$
$$3 = 3y$$
$$1 = y$$

5. $0.15 = 15¢

6. $6x^2 + 398 + 7$ has three terms. (If you combined the 398 and the 7, it would have two terms. For English majors: *The number of terms in an expression is not immutable.*)

7. {G, H} has four subsets: { }, {G}, {H}, and {G, H}.

Chapter Thirty-one
Large Numbers

The whole New New Mall lay in front of Fred. The SallyWorld sign was as tall as a two-story building. It was flashing: SallyWorld → SallyWorld → SallyWorld → SallyWorld → SallyWorld. When Fred got closer to the sign, he noticed that it was studded with diamonds—real diamonds. He also found some agate, and some albite, and some amber, and some anglesite, and some amblygonite, and some amethyst, and some andalusite, and some apatite, and some aquamarine, and some aragonite azurite, and some barite, and some benitoite, and some beryllonite, and some bixbite, and some bloodstone, and some brazilianite, and some calcite, and some carnelian, and some cassiterite, and some celestine, and some cerussite, and some chalcedony, and some chatoyant quartz, and some chrysoberyl, and some chrysocolla, and some chrysoprase, and some citrine, and some danburite, and some datolite, and some more diamond, and some diopside, and some dioptase, and some dravite, and some dumortierite, and some dunilite, and some emerald, and some enstatite, and some epidote, and some euclase, and some fire agate, and some fluorite, and some goshenite, and some garnet, and some hambergite, and some hauyne, and some heliodor, and some hematite, and some howlite, and some hypersthene, and some indicolite, and some iolite, and some jadeite, and some jasper, and some kornerupine, and some kyanite, and some labrodorite, and some lapis lazuli, and some lazulite, and some malachite, and some microcline moonstone, and some morganite, and some nephrite, and some obsidian, and some oligoclase, and some onyx, and some opal, and some orthoclase, and some pearl, and some peridot, and some petalite, and some phenakite, and some phosphophyllite, and some plasma, and some prase, and some prehnite, and some pyrite, and some rhodochrosite, and some rhodonite rubellite, and some ruby, and some rutile, and some sardonyx, and some scapolite, and some scheelite, and some sapphirine, and some schorl, and some serpentine, and some sillimanite, and some sinhalite, and some smithsonite, and some sodalite, and some spinel, and some spodumene, and some tourmaline, and some taaffeite, and some titanite, and some topaz, and some tugtupite, and some turquoise, and some tanzanite, and some vesuvianite, and some zircon.

It was not what you might call an ordinary cheap sign. (litotes)

Fred couldn't imagine how Sally got all the money. Here is the second page of the FOOL JOB application that Sally had filled out:

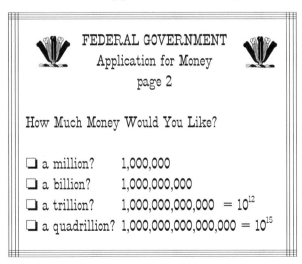

```
FEDERAL GOVERNMENT
Application for Money
page 2

How Much Money Would You Like?

❏ a million?      1,000,000
❏ a billion?      1,000,000,000
❏ a trillion?     1,000,000,000,000  = 10^12
❏ a quadrillion?  1,000,000,000,000,000 = 10^15
```

Chapter Thirty-one   Large Numbers

Sally had checked ☑ a quadrillion? $1{,}000{,}000{,}000{,}000{,}000 = 10^{15}$ because she liked words beginning with *q*. No one had ever checked the quadrillion box before. Most people asked for a billion dollars. That's one thousand million. After Sally received her government check for one quadrillion dollars, everyone's taxes went up . . . a lot.

Intermission

How much money does it take to be *infinitely* rich? A billion dollars?

First, we need to define what "infinitely rich" means. I would think that infinitely rich means that you would have enough money that you could buy anything you wanted in virtually any store you would care to shop in.

My shopping list might differ from yours. Mine might include a big new freezer and a lot of pizzas. Your list might include a mink sweater, an electric pencil sharpener and a pink sports car.

In any event, could you live on a million dollars a year? $365\overline{)1{,}000{,}000}$ = 2,739   which is $2,739/day.

For almost everyone, after they have spent $2,739 a day shopping for several weeks, they would want to stop shopping and go and play with the stuff they had purchased.

So how do you get a million dollars a year? Start with $25,000,000 and invest it at 4% interest. What is four percent of twenty-five million?

$0.04 \times 25{,}000{,}000 = \$1{,}000{,}000$ interest every year.

So, except for those (insane) people who would like to buy San Francisco, having a nest egg of twenty-five million dollars could be considered infinitely rich.

A Second Intermission

Of course, being happy can't be computed with dollar signs and percent signs.

Chapter Thirty-one   Large Numbers

With a quadrillion dollars ($1,000,000,000,000,000) Sally was able to make SallyWorld into the mall she wanted. Instead of the original 80 stores, she now had 300 stores. She kept the original gardening stores

>Abby's Agricultural Supplies
>Beatrice's Better Blooms
>Cameron's Canning Supplies
>Dakota's Digging Equipment
>Eve's Garden
>Ferdinand's Farming
>Glenda's Garden Goodies . . .

but she added 220 stores that she really liked. In one section of the mall were all the sweets:

>Patti's Pastries
>Robin's Mountain of Chocolate
>Sam's Sweet & Greasy
>Terry's Taffy
>Waddle's Doughnuts.

Fred wasn't very hungry. He still had part of a french fry in his pocket. He already knew what pastries, chocolate, taffy and doughnuts looked like. Sam's Sweet & Greasy looked a little more unusual.

Fred thought to himself *This Sam guy must have been a chemistry major. Besides selling you the sweets, he advertises the skeleton equation for how the simple sugar glucose is combined with oxygen in a cell to produce carbon dioxide and water.*

Fred dashed into the store. He wanted to meet Sam.

Chapter Thirty-one   Large Numbers

The only one in the store was the clerk who was quietly singing to herself:

♫ "Amedeo Avogadro was the coolest dude.
  ♪ He messed with moles
    ♪ And counted every little atom in a mole.
  He had their number in his hand.
  All 6.0221367×10^{23} of them."

Fred couldn't believe his eyes. This woman was tall. Really tall. Her red hair was as long as Fred was tall. When she stood near the door, Fred estimated that if she grew another three inches, her head would hit the top of the doorway.

She turned to him, and his face went the color of her hair. He couldn't speak. Fred, who had lectured for years at KITTENS University to classes that often contained hundreds of students, now was silent as melted chocolate. (simile)

"Honey, how can I help you?" she asked.

Fred stammered, "Um. Me . . . I mean I . . . could I talk to the owner, the guy Sam who owns . . . I want to see him about the chemical equation in his sign."

"I'm Sam. That's short for Samantha."

Fred's knees buckled. He passed out. We are going to have to wait a while before we balance $C_6H_{12}O_6 + O_2 \rightarrow CO_2 + H_2O$.

Your Turn to Play

1. There are a little over a hundred chemical elements. For example, carbon is a chemical element. So are iron, silver, and oxygen. Each element has been assigned an atomic weight.

For example, hydrogen (H) is 1.

Carbon (C) is 12.

Oxygen (O) is 16.

Two atoms of hydrogen can combine with one atom of oxygen and make one molecule of water (H_2O). The molecular weight of H_2O is 18 (= 1 + 1 + 16). What is the molecular weight of carbon dioxide (CO_2)?

Chapter Thirty-one   Large Numbers

2. What is the molecular weight of glucose ($C_6H_{12}O_6$)?

3. When Sam was singing, the last line of her song was **Avogadro's number**: 6.0221367×10^{23}. That's a big number.

Example: Writing out 8.3215×10^3 without the 10^3 would look like:
$$8.3215 \times 1000 = 8321.5$$

What would 6.0221367×10^{23} look like if you wrote it without the 10^{23}?

4. A **mole** is a small animal with tiny eyes that loves to dig tunnels through lawns (about 18 feet per hour for tunnels near the surface!) Make a guess how moles survive in those tunnels with very little oxygen present.

5. A **mole** is a darker colored bump on your skin. Sometimes hairs grow out of them.

   Mole sauce is not made out of moles (either the lawn-destroying animals or the skin bumps). (pronounced MOW-lay) It's made out of chocolate and chilies. They put it on chicken.

   A **mole** is a spy who works inside the enemy government. It is generally not good when you discover that your president or prime minister is a mole. (litotes)

   There is a fifth meaning of **mole**. This meaning is happier than the first four definitions. When Samantha was singing about moles, she was thinking about this fifth meaning.

   A mole of water contains 602,213,670,000,000,000,000,000 molecules of H_2O. (Recognize the number?)

   A mole of carbon dioxide contains 6.0221367×10^{23} molecules of CO_2.

   Guess how many molecules of glucose are in a mole of glucose.

6. Here is a surprise from chemistry:

   One mole of H weighs 1 gram. (See atomic weights in question 1.)
   One mole of H_2 weighs 2 grams.
   One mole of CO_2 weighs 44 grams. (Again, see question 1.)
   6.0221367×10^{23} molecules of CO_2 weighs 44 grams.

The atomic weight of calcium (Ca) is 40.
How much would 6.0221367×10^{23} atoms of Ca weigh?

167

······· COMPLETE SOLUTIONS ·······

1. The molecular weight of carbon dioxide (CO_2) = 12 + 16 + 16 = 44.

2. 6×12 + 12×1 + 6×16 = 180.

3. 6.0221367×10^{23} = 6.0221367 × 100,000,000,000,000,000,000,000
    = 602,213,670,000,000,000,000,000.

4. Moles have about twice as much hemoglobin as other mammals their size. (Hemoglobin was the "truck" that carries oxygen in the blood.)

5. A mole of any element or compound contains 6.0221367×10^{23} atoms or molecules. If I handed you a mole of salt (NaCl), you could sit down and count the number of molecules of NaCl (if you had the time and the eyesight). There would be Avogadro's number of molecules.

6. A mole of calcium would weigh 40 grams.

Chapter Thirty-two
Stoichiometry

Samantha knew what to do. She placed a box of Sam's Honey Bacon under his feet. The blood flowed to his head and his eyes popped open.

There are several possible reasons why Fred passed out. What had actually happened was that the chemical reaction $C_6H_{12}O_6 + O_2 \rightarrow CO_2 + H_2O$ hadn't been happening very much in his cells.

He had all the oxygen he needed, but his supply of glucose ($C_6H_{12}O_6$) was extremely low. He had simply forgotten to eat enough. A flea would have been starving on the amount of food Fred had eaten in the last three days.

As Fred awoke he said, "The equation on the sign on the front of your store isn't balanced."

"I know," she answered. "There wasn't any more room on the sign to put in the coefficients that would balance the equation, so I ordered a sign with just the skeleton equation."

Fred got up, and they headed to the tasting table at Sam's Sweet & Greasy. There were some paper and pencils there so that the customers could rate the various samples that Sam had put there.

Fred took one of the papers and wrote:

$$C_6H_{12}O_6 + O_2 \rightarrow CO_2 + H_2O$$

Samantha offered him one of the deep-fried sugar bits that was on the table. He took one—just to be polite—and took the tiniest bite. He put the rest in his pocket.

"You know," Fred said, "balancing chemical equations is different than solving algebra equations. With algebra equations, you combine like terms, and you move the x's to one side of the equation and the numbers to the other side. It's all very straightforward. When you balance chemical equations, you have to fool around until you get the coefficients just right."

Chapter Thirty-two   Stoichiometry

Samantha smiled. "I know. I have studied **stoichiometry**.* Shall we balance the equation?"

Fred nodded. She pushed a bowl of dried milkshake puffs toward him. He took one and put it in his pocket.

The unbalanced equation:
$$C_6H_{12}O_6 + O_2 \rightarrow CO_2 + H_2O$$

We need to find numbers (the coefficients**) to fill in the blanks:
$$\underline{}C_6H_{12}O_6 + \underline{}O_2 \rightarrow \underline{}CO_2 + \underline{}H_2O$$

When we are done, we will have the same number of C's on each side of the equation. And the same number of H's on each side. And the same number of O's on each side. Then the equation will be balanced.

Fred and Samantha looked at the original skeleton equation and noticed that $C_6H_{12}O_6 + O_2 \rightarrow CO_2 + H_2O$ had six C's on the left side and only one C on the right side.

So Sam wrote:
$$C_6H_{12}O_6 + O_2 \rightarrow 6CO_2 + H_2O$$

Balancing chemical equations is a matter of trial and error. You just fool around with the coefficients until both sides balance.

$6CO_2$ means six molecules of CO_2. Each molecule of CO_2 contains one atom of C and two atoms of O.

$6CO_2$ has six carbon atoms and twelve oxygen atoms.

---

* stoy-key-OM-eh-tree   Stoichiometry is the part of chemistry that looks at balancing chemical equations.

** A **coefficient** is the number that goes "in front."
      The coefficient of 6x is 6.
      The coefficient of $329x^2y^3$ is 329.
      The coefficient of $8H_2O$ is 8.
      The coefficient of z^6 is 1, since z^6 is the same as $1z^6$.

Chapter Thirty-two Stoichiometry

Fred and Sam looked at $C_6H_{12}O_6 + O_2 \rightarrow 6CO_2 + H_2O$ and noticed that the left side has 12 H's and the right side has 2 H's.

Sam wrote:
$$C_6H_{12}O_6 + O_2 \rightarrow 6CO_2 + 6H_2O$$

Next, they saw that the left side of the equation had 8 O's and the right side had 18 O's.

That was easy to fix:
$$C_6H_{12}O_6 + 6O_2 \rightarrow 6CO_2 + 6H_2O$$

The equation was now balanced. Both sides have 6 C's, 12 H's, and 18 O's.

Giant Hint to make stoichiometry as easy as possible: Start by balancing the letter that appears in the fewest molecules in the equation.

In $C_6H_{12}O_6 + O_2 \rightarrow CO_2 + H_2O$ you *don't* start with the oxygen. It appears in all four molecules. You start with the C or the H.

Your Turn to Play

1. Natural gas is mostly methane (CH_4). Burning natural gas in your stove or furnace gives carbon dioxide and water. The skeleton equation is $CH_4 + O_2 \rightarrow CO_2 + H_2O$. Balance that equation.
2. Hydrogen gas is H_2. Oxygen gas is O_2. When they combine, they form water. The skeleton equation is $H_2 + O_2 \rightarrow H_2O$. Balance that equation.
3. The smell of ammonia (NH_3) is hard to forget. (You'll find bottles containing ammonia in the grocery store near the bleaches and soaps.) The ammonia molecule consists of one atom of nitrogen (N), and three atoms of hydrogen.

If you combine ammonia with oxygen, you will get nitrogen gas and water. The skeleton equation is $NH_3 + O_2 \rightarrow N_2 + H_2O$. Balance it.

........COMPLETE SOLUTIONS........

1. In the original equation $CH_4 + O_2 \rightarrow CO_2 + H_2O$ the C's are already balanced.

By the **Giant Hint**, we look at the H's, which are in two of the molecules, before we look at the O's, which are in three of the molecules.

To balance the H's $\qquad CH_4 + O_2 \rightarrow CO_2 + 2H_2O$
To balance the O's $\qquad CH_4 + 2O_2 \rightarrow CO_2 + 2H_2O$

2. The original equation $\qquad H_2 + O_2 \rightarrow H_2O$
The H's are already balanced.
To balance the O's $\qquad H_2 + O_2 \rightarrow 2H_2O$
Now the H's are no longer balanced.
That's easy to fix. $\qquad 2H_2 + O_2 \rightarrow 2H_2O$

3. The original equation $\qquad NH_3 + O_2 \rightarrow N_2 + H_2O$
The **Giant Hint** doesn't help here since N is in two molecules
$\qquad\qquad\qquad\qquad\qquad\qquad\qquad$ H is in two molecules
$\qquad\qquad\qquad\qquad\qquad\qquad\qquad$ O is in two molecules

Second Giant Hint to make stoichiometry as easy as possible: Start by looking at the most complex molecules first. (We will leave the O_2 and the N_2 until last.)

Balance the N's $\qquad\qquad 2NH_3 + O_2 \rightarrow N_2 + H_2O$
Balance the H's $\qquad\qquad 2NH_3 + O_2 \rightarrow N_2 + 3H_2O$
How do you balance the O's? There are two on the left side and three on the right. Make them both into sixes. $\quad 2NH_3 + 3O_2 \rightarrow N_2 + 6H_2O$
Balance the H's again $\qquad\qquad 4NH_3 + 3O_2 \rightarrow N_2 + 6H_2O$
Balance the N's again $\qquad\qquad 4NH_3 + 3O_2 \rightarrow 2N_2 + 6H_2O$

Chapter Thirty-three
Small Numbers

Fred had taken the tiniest bite of one of the deep-fried sugar bits. Sometimes Fred kept food in his mouth for hours. He didn't want to "rush it to his tummy," as he sometimes expressed it.

Samantha's deep-fried sugar bits were loaded with sucrose. Sucrose is white table sugar. **Sucrose**, $C_{12}H_{22}O_{11}$, is a more complicated sugar that is broken down in the digestive system into the simpler sugar glucose, $C_6H_{12}O_6$, which travels in the blood to the cells.

Fred did something that should go into the record books. He managed to swallow exactly one molecule of sucrose. He certainly wasn't rushing massive amounts of food to his tummy.

How much did that micro-micro-micro piece of food weigh?

First, we convert one molecule of sucrose into moles.

Since one mole of any element or compound contains Avogadro's number of atoms or molecules (as noted five pages ago in question 5), we can use that as a conversion factor.

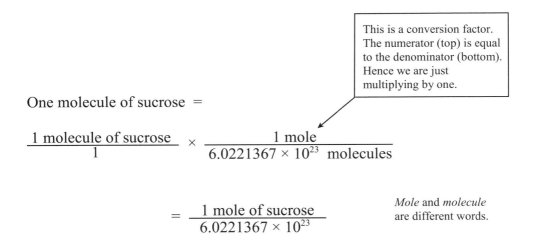

One molecule of sucrose =

$$\frac{1 \text{ molecule of sucrose}}{1} \times \frac{1 \text{ mole}}{6.0221367 \times 10^{23} \text{ molecules}}$$

This is a conversion factor. The numerator (top) is equal to the denominator (bottom). Hence we are just multiplying by one.

$$= \frac{1 \text{ mole of sucrose}}{6.0221367 \times 10^{23}}$$

Mole and *molecule* are different words.

Chapter Thirty-three   Small Numbers

Second, we convert moles of sucrose into grams.

A mole of anything weighs (in grams) the same as its atomic or molecular weight. (Seven pages ago in question 6, we noted that a mole of CO_2 weighs 44 grams, since the molecular weight of CO_2 is 44.)

The molecular weight of sucrose, $C_{12}H_{22}O_{11}$, is   *carbon* 12×12
hydrogen 22×1
oxygen 11×16

which adds up to $144 + 22 + 176 = 342$.

So one mole of sucrose weighs 342 grams. This is our second conversion factor.

$$\frac{1 \text{ mole of sucrose}}{6.0221367 \times 10^{23}} \times \frac{342 \text{ grams}}{1 \text{ mole}} = \frac{342 \text{ grams}}{6.0221367 \times 10^{23}}$$

and if you do the long division,

$$602{,}213{,}670{,}000{,}000{,}000{,}000{,}000 \overline{)342.}$$

this works out to 0.00000000000000000005679 grams.

That's one molecule of sugar. That's not exactly a ton of sugar.

Wait a minute. I, your reader, have a little question.

Yes.

How many tons is it?

How many tons is what?

You know. How many tons is 0.00000000000000000005679 grams? If you want to look at small numbers, I bet that would be pretty small.

There are 454 grams in a pound and 2000 pounds in a ton. That gives us the two conversion factors.

$$\frac{0.00000000000000000005679 \text{ grams}}{1} \times \frac{1 \text{ pound}}{454 \text{ grams}} \times \frac{1 \text{ ton}}{2000 \text{ pounds}}$$

$= 0.00000000000000000000000000062544$ tons.

Chapter Thirty-three   Small Numbers

Scientists who have to use numbers like 0.0000000000000000000000000062544 in their everyday work would go crazy counting zeros all day long. It might be simpler to write $\dfrac{6.2544}{10{,}000{,}000{,}000{,}000{,}000{,}000{,}000{,}000}$

or even simpler to write $\dfrac{6.2544}{10^{28}}$

Your Turn to Play

1. Please don't look at the answer before you have given this some thought. It will spoil the fun. Your question: *In your everyday life, you often use a number that is even smaller than* $\dfrac{6.2544}{10^{28}}$ *What is that number?*

2. The chemical formula for butane is C_4H_{10}. If you use a butane lighter to light your barbeque, the skeleton equation is $C_4H_{10} + O_2 \rightarrow CO_2 + H_2O$. Balance that equation.

3. Solve $23x + 20 - 5x = 110$

4. Here is a maze with two solutions. Draw one that has less than $\dfrac{6.2544}{10^{28}}$ solutions.

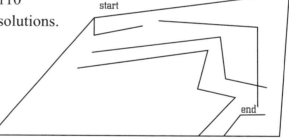

. . . . **COMPLETE SOLUTIONS** . . . .

1. Tell me, how many times last year did you sing the song:
   Some moles like to dig,
   Others grow a hair.
   Some will be a spy.
   While the mole I love
   Is the one from chem
   That's 6.0221367×10^{23}.

The number of times you sang that song last year is a number that is smaller than $\dfrac{6.2544}{10^{28}}$ It is zero.

2. Original equation $C_4H_{10} + O_2 \rightarrow CO_2 + H_2O$
Balance the C's $C_4H_{10} + O_2 \rightarrow 4CO_2 + H_2O$
Balance the H's $C_4H_{10} + O_2 \rightarrow 4CO_2 + 5H_2O$

Now, we are a bit stuck. There are 13 O's on the right side. There are two different approaches you may have used.

First approach: I need to have an even number of O's on the right side in order to put a coefficient in front of the O_2 on the left side. So I'll double all the coefficients. $2C_4H_{10} + 2O_2 \rightarrow 8CO_2 + 10H_2O$
Balance the O's $2C_4H_{10} + 13O_2 \rightarrow 8CO_2 + 10H_2O$

Second possible approach: I know that the coefficients in a chemical equation have to be whole numbers.* You can't have fractional parts of a molecule.
Ignoring that fact,
I will balance the O's $C_4H_{10} + 6½O_2 \rightarrow 4CO_2 + 5H_2O$
Then double the equation
to eliminate the fractions $2C_4H_{10} + 13O_2 \rightarrow 8CO_2 + 10H_2O$

3. $23x + 20 - 5x = 110$
   $18x + 20 = 110$ Combine like terms
   $18x = 90$ Subtract 20 from both sides
   $x = 5$ Divide both sides by 18

4.

∗ The **whole numbers** are {0, 1, 2, 3, . . .}.

Chapter Thirty-four
Shortcuts

Samantha offered Fred a buttered lollipop. She told him that buttered lollipops were one of her best sellers. He thanked her and put it in his pocket "for later."

As everyone knows who has ever carried a cube of butter in their pocket, Fred had made a mistake. The deep-fried sugar bits, the dried milkshake puffs, the french fry, and all the other items of food that Fred had placed in his pocket over the last several weeks—they all stuck together. It would only be later when he reached into his pocket for his keys that he would find that his hand was stuck in his pocket.

Fred thanked Samantha and headed outside to see the rest of the mall. He was thinking about the name she had given her store, Sam's Sweet & Greasy. That really said it all. It was truth in advertising. When he looked at the signs of the other stores, he started to see a pattern.

Patti's Pastries
   Robin's Mountain of Chocolate
      Terry's Taffy
         Waddle's Doughnuts
            Harry's Hamburgers & Shakes
         Sally's Sugar Buns
            Ivy's Ice Cream
               Peter's Pies

If your store offered sugar and grease, it was a success. Why? You don't see a lot of stores like   Paul's Potatoes
                  Carl's Carrots
               Larry's Lettuce
            Tabitha's Tomatoes

The answer lies in our biology. What things in nature are sweet? The ripe fruits. Sweetness rings a bell inside our heads. It says, "I'm ready to be eaten."

What about greasy things? Why are foods with a lot of fat in them so appealing? The answer is that the greasy foods are labor-saving. Do you ever see lions eating grass? No. They lie around most of the day playing video games, while the antelopes have to spend their whole day grazing on plant food. The antelopes have to do a lot of eating because grass contains very few calories.

Lions take a shortcut and have antelope burgers. It's fast food.

Lions love to BBQ!

> *Intermission*
>
> Now you know why some people become bank robbers, car thieves, and shoplifters. It's a shortcut.

Life offers lots of shortcuts. A shortcut is a quick grab at something pleasurable. Invariably, it is followed by long periods of regret or pain.

▷▷▷ **Multiple Choice Quiz** ◁◁◁

1. Which of the following is true?
   - A) The future is coming.
   - B) It isn't.

Is that a dumb quiz? But how many people do you know who act as if the correct answer is B? There are plenty of examples:

Example #1: Since this is a biology book, we have to start with sucking on cigarettes. It's a shortcut. Immediate pleasure. *And they pretend the future doesn't exist.* At your favorite office supplies store, notice the employees standing out in front sucking on their cigarettes. In the winter when it's snowing, it's not a pretty sight.

"Smoking is the major cause of lung cancer, which is now the leading cause of death among women." (Cecie Starr, *Biology: Concepts and Applications*, p. 506)

Chapter Thirty-four Shortcuts

Example #2: A diet of chocolate ice cream, fast foods, sugar drinks, fatty meats, and candy. Immediate pleasure. *And they pretend the future doesn't exist.*

Example #3: In working the problems in the *Your Turn to Play*, the quick and easy approach is to read the problem and then just read the answer, rather than working it out for yourself. Immediate pleasure. *And they pretend the future doesn't exist.* The future does happen when you reach the Bridge.

Example #4: Lying. Immediate pleasure (or the avoidance of pain). *And they pretend the future doesn't exist.* One of the biggest pleasures in life is human relationships—and lying destroys relationships.

   I bet you are happy with this list of examples, *except for the ones that applied to you.*
   I, your reader, have to agree. How did you know?
   You are human, aren't you? The old saying is that those who don't drink are happy to hear speeches about the evils of alcohol abuse.

Your Turn to Play

1. List five other shortcuts that people take.
2. $\frac{4}{5} + \frac{3}{7}$
3. $\frac{4}{5} - \frac{1}{3}$
4. $\frac{6}{7} \times \frac{1}{8}$
5. $\frac{2}{3} \div \frac{4}{9}$
6. 2 is what percent of 8?
7. Express $\frac{3}{4}$ as a percent.

Chapter Thirty-four   Shortcuts

........COMPLETE SOLUTIONS........

1. Your list may differ from mine.
    1. Jaywalking as a shortcut to getting across the street faster.
    2. Hitting someone as a shortcut to working things out peacefully.
    3. Cheating on an exam as a shortcut to learning the material.
    4. Cheating on your spouse.
    5. Loafing around and avoiding exercise.
    6. Reading "trash" and avoiding reading the "good stuff."
    7. Letting messes accumulate in your room.
    8. Letting the lawn grow four feet tall.
    9. Not changing the oil in your car until the engine starts to squeak.
    10. Not flossing your teeth.
    11. Not making a personal list of shortcuts that you would like to eliminate from your life.

2. $\frac{4}{5} + \frac{3}{7} = \frac{28}{35} + \frac{15}{35} = \frac{43}{35} = 1\frac{8}{35}$

3. $\frac{4}{5} - \frac{1}{3} = \frac{12}{15} - \frac{5}{15} = \frac{7}{15}$

4. $\frac{6}{7} \times \frac{1}{8} = \frac{\cancel{6}^{3}}{7} \times \frac{1}{\cancel{8}_{4}} = \frac{3}{28}$

5. $\frac{2}{3} \div \frac{4}{9} = \frac{2}{3} \times \frac{9}{4} = \frac{\cancel{2}^{1}}{\cancel{3}_{1}} \times \frac{\cancel{9}^{3}}{\cancel{4}_{2}} = \frac{3}{2} = 1\frac{1}{2}$

6. You divide the number closest to the *of* into the other number. So *2 is what percent of 8?* becomes $\frac{2}{8}$ which is $\frac{1}{4}$ which is 25%.

7. The conversion of $\frac{3}{4}$ into a percent is one of the Nine Conversions you memorized in *Life of Fred: Decimals and Percents*.

$\frac{1}{2} = 50\%$ $\frac{1}{4} = 25\%$ $\frac{3}{8} = 37½\%$

$\frac{1}{3} = 33⅓\%$ $\frac{3}{4} = 75\%$ $\frac{5}{8} = 62½\%$

$\frac{2}{3} = 66⅔\%$ $\frac{1}{8} = 12½\%$ $\frac{7}{8} = 87½\%$

Chapter Thirty-five
Reducing Fractions

Fred knew about two parts of the **SallyWorld** mall. He had seen the agricultural section earlier in the day, and he had just been in the food section. After Sally had become owner, she had added 220 stores to the mall. Since she was only two years old, Sally didn't include any auto parts stores in her mall. There were no stores devoted to hunting and fishing. It was no surprise that the biggest section of **SallyWorld** was toys. And most of the stores were painted either pink or purple—Sally's favorite colors.

He glanced inside the doll store. There were hundreds of pink boxes of dolls. The sign in the window read:

He couldn't figure it out. When he had owned the mall, he was going broke. When Sally owned it, there was no shortage of money. He didn't know about the quadrillion dollars that she had received from the Federal government. She had spent only a billion dollars to build **SallyWorld**, so she had

$$\begin{array}{r} 1{,}000{,}000{,}000{,}000{,}000 \\ - \phantom{000{,}000{,}}1{,}000{,}000{,}000 \\ \hline 999{,}999{,}000{,}000{,}000 \end{array}$$

in spare change. She could afford to give away dolls and dresses.

The nine hundred ninety-nine trillion, nine hundred ninety-nine billion dollars that Sally had deposited in KITTENS Bank wasn't an everyday deposit. (litotes)

How much did the Federal government's transfer of a quadrillion dollars to Sally cost each taxpayer? Suppose there are, say, two hundred million taxpayers. Do we add, subtract, multiply, or divide? The numbers are too big to think about.

We can use the **General Rule**: *If you don't know whether to add, subtract, multiply or divide, first restate the problem with really simple numbers.*

Suppose the government had spent $12, and there were 4 taxpayers. Then we know that each person would have to pay $3 in taxes. How did we get the $3? We divided.

So, in the original problem, one quadrillion dollars that is paid for by two hundred million taxpayers would be:

$$\frac{\$1,000,000,000,000,000}{200,000,000}$$

Do you remember how to reduce fractions? If we started with a fraction like $\frac{24}{36}$ we would look for a number that could divide evenly into both the top and the bottom. If, for example, we divided top and bottom by 6 we would get $\frac{24}{36} = \frac{4}{6}$

We could further reduce $\frac{4}{6}$ by dividing numerator and denominator by 2: $\frac{4}{6} = \frac{2}{3}$

We could take the giant fraction $\frac{\$1,000,000,000,000,000}{200,000,000}$ and divide top and bottom by 10. That would strike off a zero from the top and the bottom.

$$\frac{\$1,000,000,000,000,00\cancel{0}}{200,000,00\cancel{0}}$$

Hey! That was fun. Let's do it a lot.

$$\frac{\$1,000,000,0\cancel{00},\cancel{000},\cancel{000}}{2\cancel{00},\cancel{000},\cancel{000}}$$

That leaves us with $\frac{\$10,000,000}{2}$ which equals $5,000,000.

So each taxpayer (if everyone paid the same tax) would receive a bill from the federal government for five million dollars.

Chapter Thirty-five Reducing Fractions

Fractions like $\frac{300}{5770}$ are easy to reduce. You can divide numerator and denominator by 10. $\frac{30\cancel{0}}{577\cancel{0}} = \frac{30}{577}$

Sometimes it is harder to see what number will divide evenly into both the top and bottom of a fraction.

How might you reduce $\frac{9723960775}{11398267400}$?

> Rule: If the last digit is 0 or 5, it's divisible by 5.

How might you reduce $\frac{3677554}{967063336}$?

> Rule: If the last digit is 0, 2, 4, 6, or 8, it's divisible by 2.

How might you reduce $\frac{200007}{10000002}$?

> Rule: If the sum of the digits is divisible by 3, so is the number.

Your Turn to Play

1. There are 6,012 dolls in the store. Could two kids share them equally?
2. Could five kids share 6,012 dolls equally?
3. Could three kids share 6,012 dolls equally?
4. $\frac{26}{65}$ can be reduced. None of the three rules given above apply in this case. Reduce $\frac{26}{65}$
5. $\frac{5}{22} + \frac{6}{22}$

183

Chapter Thirty-five Reducing Fractions

······· COMPLETE SOLUTIONS ·······

1. 6,012 is an even number. Its last digit is 0, 2, 4, 6, or 8. Therefore, it is divisible by 2.

2. The last digit of 6,012 is not 0 or 5. It is not divisible by 5.

3. The sum of the digits of 6,012 (6 + 0 + 1 + 2) is equal to 9. Since the sum of the digits is divisible by 3, we can say that 6,012 is divisible by 3.

4. To reduce $\frac{26}{65}$ you have to find a number that divides evenly into both 26 and 65.

✓ Since they are not both even, 2 doesn't divide into them both.
✓ 3 doesn't divide evenly into 26.
✓ 5 doesn't divide into 26.
✓ 7 doesn't divide into 26.
✓ If 3 doesn't divide into 26, then 9 won't.
✓ 10 doesn't divide into 26.
✓ 11 doesn't divide into 26.
✓ 12 doesn't divide into 26.
✓ 13 will go into 26. Will it also go into 65? Let's try: $13\overline{)65}$
  $\underline{-65}$
  0

So $\frac{26}{65} = \frac{2}{5}$

5. $\frac{5}{22} + \frac{6}{22} = \frac{11}{22}$ and by the General Rules, we reduce $\frac{11}{22}$ to $\frac{1}{2}$

═══

Here are the four General Rules (from *Life of Fred: Fractions*):

(1) Reduce fractions in your answers as much as possible.

(2) Fractions like $\frac{0}{4}$ are equal to 0.

(3) Fractions like $\frac{4}{4}$ are equal to 1.

(4) Division by zero is not permitted.

Chapter Thirty-six
Division by Zero

Fred walked by the doll sto . . .

Hold it! Wait a minute. Before you start this chapter, I, your reader, need to get something straightened out. At the end of the previous chapter you laid out the four General Rules for fractions.

I'm happy with Rule #1: Reduce fractions in the answers.
I'm happy with Rule #2: Fractions like 0/4 are equal to 0.
I'm happy with Rule #3: Fractions like 4/4 are equal to 1.

But who made you dictator? What gives you the right to say that division by zero is not permitted?

I'm not trying to tell you frogs eat antelopes. There are lots of good arguments why division by zero is not permitted.

Gimme one.

Here's one.

You say you are happy with General Rule #2, which states that any fraction that has a numerator equal to zero is equal to zero. Right?
Yup.
So $\frac{0}{0}$ must equal 0.
Yup.

You say you are happy with General Rule #3, which states that any fraction whose numerator equals its denominator must be equal to one. Right?
Yup.
So $\frac{0}{0}$ must equal 1.
Yup.

If $\frac{0}{0} = 0$ and $\frac{0}{0} = 1$, then it must be true that $0 = 1$.

185

So the minute you allow division by zero, you are forced to admit that 0 = 1.

This is called an **argument by contradiction**. In the proofs in geometry, arguments by contradiction are used frequently. These proofs are called **indirect proofs**.

In an indirect proof, you assume the opposite of what you want to show is true. Then you argue that under that assumption you will arrive at nonsense.

I assumed that division of zero was legal. I then showed that on the basis of that assumption we have 0 = 1. Therefore, the assumption must be false.

Okay. I'll buy the fact that 0/0 doesn't work. But what about something like 5/0. What's wrong with that?

I claim that $\frac{5}{0}$ doesn't make sense either.

Okay. Prove it.

I'm going to need a little fact from algebra, and this is only a pre-algebra book.

How little is "little"?

It's not very big. It's called **cross multiplying**. When you have two fractions equal to each other, you can cross multiply. Suppose, for example, you have

$$\frac{2}{3} = \frac{6}{9}$$

We know those fractions are equal since $\frac{6}{9}$ reduces to $\frac{2}{3}$

If we cross multiply, we get $2 \times 9 = 3 \times 6$.

In general, if we have any four numbers, a, b, c, d, and we know that $\frac{a}{b} = \frac{c}{d}$ we can say that ad = bc.

Here's why it's called cross multiplying: $\frac{a}{b} \times \frac{c}{d}$

implies $ad = bc$.

Chapter Thirty-six   Division by Zero

Examples of cross multiplying

example #1: If $\frac{4}{5} = \frac{80}{100}$ then $4 \times 100 = 5 \times 80$

example #2: If $\frac{2}{x} = \frac{6}{15}$ then $(2)(15) = 6x$

example #3: If $\frac{21}{12} = \frac{y}{4}$ then $(21)(4) = 12y$

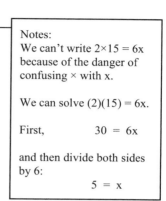

Note:
We can solve $(21)(4) = 12y$.

First, $84 = 12y$

and then divide both sides by 12:
          $7 = y$

     Now I can prove that $\frac{5}{0}$ doesn't make sense. I'll use an indirect proof.

     Start by assuming that $\frac{5}{0}$ does make sense. Suppose that it is equal to some number c.

Then $\frac{5}{0} = c$.

Then $\frac{5}{0} = \frac{c}{1}$ since $c = \frac{c}{1}$

Then $(5)(1) = 0c$ by cross multiplying

Then $5 = 0c$

Then $5 = 0$ since zero times anything equals zero

We have arrived at nonsense. (In math, it's called a contradiction.) Therefore, our assumption that $\frac{5}{0}$ does make sense must be false. In other words, you can't divide by zero.

Chapter Thirty-six   Division by Zero

Your Turn to Play

1. Solve $\dfrac{x}{5} = \dfrac{24}{15}$

2. If the government spends a trillion dollars, and there are two hundred million taxpayers, how much will the average tax bill be?

3. Name the eight subsets of {a, b, c}.

4. Change $7\dfrac{3}{8}$ into an improper fraction.

·······**COMPLETE SOLUTIONS**·······

1. Original equation $\qquad\qquad \dfrac{x}{5} = \dfrac{24}{15}$

   Cross multiply $\qquad\qquad$ 15x = 5(24)
   Arithmetic $\qquad\qquad\qquad$ 15x = 120
   Divide both sides by 15 $\quad$ x = 8

2. $\dfrac{\$1,000,000,000,000}{200,000,000} = \dfrac{\$1,000,0\!\!\!\not0\!\!\!\not0,0\!\!\!\not0\!\!\!\not0,0\!\!\!\not0\!\!\!\not0}{20\!\!\!\not0,0\!\!\!\not0\!\!\!\not0,0\!\!\!\not0\!\!\!\not0} = \dfrac{\$10,000}{2} = \$5,000$

3. { }, {a}, {b}, {c}, {a, b}, {a, c}, {b, c}, {a, b, c}.

4. $7\dfrac{3}{8} = \dfrac{59}{8}$

Hey! You never proved that cross multiplying is true!

That's super easy. Start with $\qquad\qquad \dfrac{a}{b} = \dfrac{c}{d}$

Multiply both sides by $\dfrac{bd}{1}$ $\qquad \dfrac{a}{b}\dfrac{bd}{1} = \dfrac{c}{d}\dfrac{bd}{1}$

And then cancel $\qquad\qquad\qquad \dfrac{a}{\cancel{b}}\dfrac{\cancel{b}d}{1} = \dfrac{c}{\cancel{d}}\dfrac{b\cancel{d}}{1}$

$\qquad\qquad\qquad\qquad\qquad\qquad$ ad = cb

The Bridge

from Chapters 1 to Chapter 36

first try

Goal: Get 9 or more right and you cross the bridge.

1. The phone rang. Darlene looked at the clock. It was 2 a.m. She knew who was calling.

"I got to tell you something," Joe began. "I've decided what I want to be after college. I've got it all figured out."

Darlene put her head on her pillow and held the phone to her ear. She mumbled, "Joe, I was sleeping."

"That's funny," Joe said. "So was I. Then I got this great idea. You remember when we went to the movies and saw those guys riding around in armor and carrying swords?"

"Joe, that was yesterday evening. I remember," Darlene said.

"Well, I want to be one of them. I just can't remember what those guys were called."

Darlene said, "They are knights."

"And you can polish my armor," Joe said.

"Good night Joe," Darlene said as she hung up the phone.

Joe liked that. He would be called Good Knight.

Iron armor rusts. Iron (Fe) combines with oxygen (O_2) to form rust (Fe_2O_3). Balance that equation.

2. There were 20 knights in the movie. (Joe had counted them.) Three of them were bad knights. What percentage of the knights were bad?

3. In the movie the three bad knights stole 4,000,700 gold coins. (Joe counted them.) Name the smallest number that is larger than 4,000,700 so that the three knights could have equally divided the coins among themselves.

4. In movies involving knights there are always 5 fighting scenes for every 2 scenes that involve kissing. If you made a movie with 18 kissing scenes, how many fighting scenes would it have?

In other words, solve $\frac{5}{2} = \frac{x}{18}$

The Bridge
from Chapters 1 to Chapter 36

5. If Joe's armor weighed 40 pounds and a 2 ounce bolt fell off, how much would it now weigh?

6. The big fighting scene at the end of the movie took 12 minutes. (Joe timed it.) How long is that in hours?

7. Which of these is largest? 999, $10 \div 0.01$, or π^4

8. Because Joe eats so much popcorn at the movies, they have given him a special discount card.

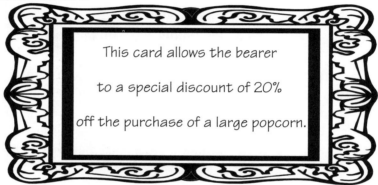

This card allows the bearer to a special discount of 20% off the purchase of a large popcorn.

A large popcorn normally cost $7. How much would Joe have to pay?

9. $\frac{3}{4} - \frac{3}{5}$

10. $\frac{3}{4} \div \frac{3}{5}$

The Bridge
from Chapters 1 to Chapter 36

second try

---

1. Joe lay in bed for an hour thinking about himself as the Good Knight. He knew that the first thing about being a knight was getting some armor. He said to himself, "A knight without armor would be like cheese without mice." (Joe wasn't very good with similes. "Ice cream without a spoon" or "Peanut butter without jelly" would have been a little better.)

 At 3 a.m. Joe leaped out of bed singing a song from the movie he and Darlene had seen last night. The only words he could remember from the song were *knight* and *unwinceable*.* Since there were 50 different words in that song, what percentage of the words could Joe remember?

2. Joe tried a different song from the movie, but could only remember two words from that song also: "Say Ma!" He couldn't figure that out. He substituted, "Say Cheese!" and it made more sense. Besides, he called his mother "Mama," not "Ma."

 The tenants on the floor above Joe were banging on Joe's ceiling. They banged on their floor three times for every two words that Joe sang. (Joe counted.) Joe wondered how many words he would have to sing in order to have them bang 39 times. In other words, solve $\frac{3}{2} = \frac{39}{x}$

3. Joe headed to the KITTENS University phonebook to see if he could find some place that would sell him armor at three in the morning. He looked under *armor* and found nothing. Then under *clothing for knights* and found nothing. He then turned to the beginning of the pages that were colored yellow and began to read. He read 1.6 pages per minute. When the sun came up at 6 a.m., how many pages had he read? (Note: He started reading at 3 a.m., so he had read for 180 minutes.)

4. Joe had gotten as far as *igloo repair* in the phone book and was starting to get hungry. He walked toward the kitchen at 2.5 feet per second. His kitchen was 20 feet away. How long did it take him to get to the kitchen? (One approach to this problem is to use a conversion factor.)

---

* Movie buffs now can guess what movie Joe and Darlene went to see.

The Bridge
from Chapters 1 to Chapter 36

5. In the kitchen Joe exclaimed, "I know! I can use aluminum foil for my armor." He tore off a length of 27 feet from the roll. If the roll is 1.1 feet wide, what is the area he tore off?

6. The atomic weight of aluminum (Al) is 27. Joe's piece of aluminum weighed 54 grams. How many moles of aluminum did Joe have?

7. How many atoms are in one mole of aluminum?

8. Joe managed to cut himself on the aluminum foil. He headed to the bathroom and found some hydrogen peroxide (H_2O_2) which is an antiseptic (= kills germs). Hydrogen peroxide decomposes into water and oxygen. The skeleton equation is $H_2O_2 \rightarrow H_2O + O_2$. Balance that equation.

9. $(3\frac{1}{3})^3$ Do not leave your answer as an improper fraction.

10. Solve $5x + 13 + 7x = 16x + 1$

The Bridge
from Chapters 1 to Chapter 36

third try

1. Joe had the large bottle of hydrogen peroxide. He had purchased it at a hospital supply store. The bottle was 18" high and had a radius of 4". Using $\pi = 3$, find its volume.

2. Joe used 32.7% of the bottle on his cut. What percent did he leave in the bottle?

3. How much would 4 moles of H_2O_2 weigh? (Atomic weights of H and O are 1 and 16.)

4. If the sum of the digits of a number is divisible by 7, is the number divisible by 7?

5. Back in the kitchen, he wrapped the aluminum foil around his chest. The foil was 27 feet long. How many inches is that?

6. The foil went ten times around his chest. What is Joe's chest measurement? Round your answer to the nearest inch.

7. Joe wondered what a knight would eat for breakfast. To him, the answer was obvious: rocky road ice cream. "After all," Joe said to himself aloud, "knights have a rocky road to travel as they go around killing things." Darlene never talked about the joys of going around and killing things. This was perhaps because she had a lot less of the hormone $C_{19}H_{28}O_2$ (testosterone) running around in her body. What is the molecular weight of testosterone? (Atomic weight of carbon is 12.)

8. Joe accidentally dribbled a little rocky road ice cream onto his chest. He looked at the nuts and marshmallows in the ice cream on the aluminum foil. "It looks like the medal they give you when you are wounded," Joe said. (Joe did a lot of his thinking aloud.) He dribbled 7 medals on his chest. Each weighed 8.9 grams. What was the weight of his medals?

9. There was a knock on the door. Joe's neighbor from upstairs had come down to talk to Joe about his singing during the night. He was going to explain to Joe that losing three-eighths of a night's sleep was unacceptable. What percentage is three-eighths?

10. The neighbor pounded on Joe's door. When Joe opened the door, the neighbor was thunderstruck. There was Joe in his pajamas with aluminum foil wrapped around his chest with gobs of ice cream on the foil.

193

The Bridge
from Chapters 1 to Chapter 36

When the neighbor couldn't speak, Joe broke into his "Say Cheese!" song.

The neighbor slowly backed away and headed back up the stairs. When he got back to his apartment, he told his wife, "We got a real nut case downstairs. He's one sandwich short of a picnic." (metaphor)

Joe couldn't figure out why the neighbor had knocked on his door and then hadn't said anything. Joe said aloud, "He must be crazy or something."

Joe sat down and put $3\frac{1}{4}$ pounds of ice cream in his bowl and then put $1\frac{3}{4}$ pounds of ice cream in a second bowl "for his horse." How much more was in Joe's bowl than the horse's?

The Bridge
from Chapters 1 to Chapter 36

fourth try

1. Since Joe didn't have a horse yet, he ate both the $3\frac{1}{4}$ pounds of ice cream in his own bowl and the $1\frac{3}{4}$ pounds of ice cream in his horse's bowl. That's 5 pounds of ice cream. How many ounces is that? (1 lb. = 16 oz.)

2. After his breakfast of ice cream, Joe figured he should have some dessert before he headed out to do his work as a knight. Frozen Sluice was the perfect dessert. It was almost 100% sugar. In fact, it was exactly 99.077% sugar. What percentage wasn't sugar?

3. Joe had made frozen Sluice by pouring Sluice into ice cream cones and then putting them into the freezer. The cones were 4" high and had a radius of 1.5". Using $\pi = 3$, find the volume of those Sluice desserts.

4. "A bit of bacon wouldn't hurt," Joe said aloud. "After all, it may not be until noon before my next meal." Tristearin ($C_{57}H_{110}O_6$) is a typical animal fat. "I'll just have a pound of bacon. I don't want to overdo it." Here is the skeleton equation of how tristearin is metabolized at the cellular level:

$$C_{57}H_{110}O_6 + O_2 \rightarrow CO_2 + H_2O \quad \text{Balance it.}$$

5. "I gotta give my stomach a little time to rest before I head on out." Joe's stomach hurt only a little bit. He lay down and was soon asleep. Joe snored at the rate of 6 snores every 13 seconds. In 91 seconds, how many snores would Joe emit? (Hint: *6 snores every 13 seconds* can be used as a conversion factor.)

6. The whole numbers are {0, 1, 2, 3, . . . }. If x is a whole number, how many different solutions would x < 7 have? (Count carefully!)

7. When Joe awoke, it was noon. He headed back into the kitchen for lunch. At the rate Joe was eating, he would consume a million calories in a month's time. If a month has 30 days, how many calories per day would he eat? Round your answer to the nearest calorie.

8. Solve $30 + 6x = x + 85$

9. $39.7 + 0.08$

10. $(2\frac{1}{5})^2$ Do not leave your answer as an improper fraction.

The Bridge
from Chapters 1 to Chapter 36
fifth try

1. Joe packed an afternoon snack (8,000 calories) in his backpack: some cheeseburgers and a quart of Sluice. The Sluice was 27% of the total calories in his afternoon snack. How many calories were in the Sluice?

2. Joe thought it seemed awfully late in the day to start his new life as a knight. We know that because Joe said aloud, "It seems awfully late in the day for me to start my new life as a knight." (Darlene always knew what was on Joe's mind. All she had to do was listen.)

    Joe headed over to Darlene's apartment and knocked on the door. The door was 6'8" high and 36" wide. What is the area of that door?

3. The area was important because a painter was busy painting that door a bright pink. (Darlene's favorite color.) A gallon of pink paint will cover 300 square feet. How many square feet will a pint cover?
(1 gallon = 4 quarts. 1 quart = 2 pints.)

4. Joe's hand had a bit of bright pink paint on it. When Joe had knocked on the door, Darlene had been working on painting her fingernails and her toenails. She had completed 4 of her nails. What percentage of the job had she completed?

5. When Joe entered her apartment, he was still dressed in his aluminum foil armor with the gobs of ice cream on his chest. Darlene reached for an Alka-Seltzer® tablet and dropped it into a glass of water. The sodium bicarbonate and the citric acid in the tablet react:

$$NaHCO_3 + H_3C_6H_5O_7 \rightarrow CO_2 + H_2O + Na_3C_6H_5O_7$$

(The fizz comes from the carbon dioxide.) Balance that equation. (Na is the chemical symbol for sodium, because S is the symbol for sulfur.)

6. $7^3 \times 12^5 \times \pi \times 0 = ?$

7. "I need a horse," Joe announced. Joe had skipped the usual pleasantries such as, "Hello Darlene," or "Oh, have I come at a bad time?"

    Darlene was almost used to Joe's gaucheries. (GO-sha-REES = lacking in the social graces) She responded, "I don't have any. I'm fresh out." She went back to painting her nails.

The Bridge
from Chapters 1 to Chapter 36

"Oh," said Joe. He had known Darlene for years, but the fact that she didn't own a horse came as a little surprise to Joe.

"Does your lease allow you to have a horse?" Darlene asked.

Joe knew all about the lease's provision for animals. Last year when he had decided to get some ducks, he read in the lease: Animals over five pounds require a deposit. The deposit is $1.20 for each pound of the animal's weight.

If a horse weighs a ton (= 2000 lbs.), how much will the deposit be?

8. Joe decided that he would be a knight without a horse. "But I do need a sword," Joe said. "Every knight in the movie had a sword." Darlene pointed to the kitchen. Joe headed into the kitchen and came back with Darlene's electric knife. He plugged it in and tried it out on one of his cheeseburgers. The cheeseburger originally weighed 2 pounds. He cut off an eleven-ounce portion and ate it. How much of the cheeseburger was left?

9. To "wash down the cheeseburger" (as Joe often expressed it) he drank 4 ounces of the quart of Sluice that he had brought with him. What percentage of the quart did he drink? (1 quart = 32 ounces)

10. Darlene explained to Joe that in order to become a knight you need to go through some ceremony in which a king declares you to be a knight.

"Being a knight is all too complicated," Joe complained. "After I graduate from KITTENS University, I think I will be elected king instead."

Darlene noticed that when Joe had cut his cheeseburger, it had been sitting on top of her *Bridal Hairdos* magazine. He had cut through 16 of the 256 pages of the magazine. What fraction of the magazine had he cut through? Reduce your answer as much as possible, which is the first of the four General Rules for fractions.

Chapter Thirty-seven
Bones

Fred walked by the doll store. He didn't need any more dolls. In his office he had Kingie—the doll he had received at the King of French Fries store when he was four or five days old. The man at King of French Fries had given the doll to him for free because he had felt sorry for the kid. It was a good boy's doll. It had a beanbag base so that it wouldn't fall over when you set it down. It also played the K.F.F. theme song when you squeezed its tummy: "Butter fries are butter! ♪♫. You wouldn't want anything utter."

Kingie

It happened in an instant.

Things happen in life like that—just when you least expect them. Sometimes it's good; sometimes it's bad.

There are a thousand examples of times when life turns a corner: (metaphor)

✓ You're driving, and your attention wanders as you adjust the radio.

✓ You suddenly discover that you are not the center of the universe. There are other human beings out there that have feelings and needs just like you have. This can occur in individuals as early as four or five years old. Some people never figure that out. In philosophy, this condition is called **solipsism**—"Only I really exist." [SAUL-ip-SEZ-em]

✓ You suddenly realize what is the central major interest in your life: bowling or pizza or mathematics or real estate or ballet or God.

✓ Your doctor is reading the results of your heart test. He involuntarily says, "Oh no!" He sets down the paper, screws the cap back on his fountain pen, and looks straight at you. He begins, "I'm sorry to say. . . ." (The central major interest in your life may change at this point.)

Chapter Thirty-seven   Bones

For Fred, he was about to learn a lot about bones—his bones. As he was thinking about Kingie, he walked off the curb. The curb was only four inches high, but it was enough to twist his ankle and send pain radiating up his leg.

As he dropped to the floor, he tumbled down the escalator hitting the corners of every one of the steel steps. Getting hit with a steel baseball bat is generally considered preferable. Bats are round and not sharp-edged.

A crowd gathered around him. Several people whipped out their cell phones . . . and took pictures.

One bystander commented, "I guess he won't be spending a long time at **SallyWorld** today."

Half-conscious, Fred thought that the bystander had said, "I guess he's not long for this world." Fred thought to himself *I'm ready. I'm glad I paid attention in Sunday School.*

A man from security arrived. He held the crowds back so no one would come and try to pick Fred up. Fractured bones can have sharp ends that can cut into nearby muscles, nerves or blood vessels.

He called the captain of security,
      who called the management office,
           who called the owner of the mall—Sally.

Sally took the call on her pink cell phone. When she heard what had happened, she dialed 9-1-1 and told them to come and "scoop up someone who had fallen down the escalator."

When Fred's eyes opened, he could feel pain in all parts of his body. *I'm pretty sure this isn't Heaven.* One bystander offered him a slurp of his 96-ounce Sluice drink. Another offered him a bite of his cheeseburger.* Fred declined both of these offers.

It took twenty minutes for the medics to arrive. All they had been told is that the victim was at the foot of the escalator in **SallyWorld**. They had never heard of **SallyWorld**, which was understandable since it

---

★ Can you guess why you don't give food or drink to someone with a possible bone fracture? That's an advanced first aid question. I'll give the answer at the end of this chapter after the *Your Turn to Play*.

Chapter Thirty-seven   Bones

had come into existence only hours ago. And once they found the mall, they didn't know which of the fifteen escalators was the right one.

Meanwhile, Fred didn't move an inch. He couldn't. It hurt too much. The crowd slowly melted away. (metaphor) They got bored watching nothing happen.

"He's so small," one medic said to the other.

"And so messed up," the other replied.

They wondered how they could lift Fred up and put him on the stretcher. One medic ran into Carol's Cooking for Kids store and brought out a spatula. They slid it under Fred and moved him the same way you might move a fried egg from a frying pan to a plate.

a Fred scooper

At the hospital they moved Fred off to the X-ray department in order to determine whether any bones were broken. Fred's body was covered with purple marks that looked like ▦▦▦▦▦. They were made by the edges of the escalator steps. They did a full-body X-ray.

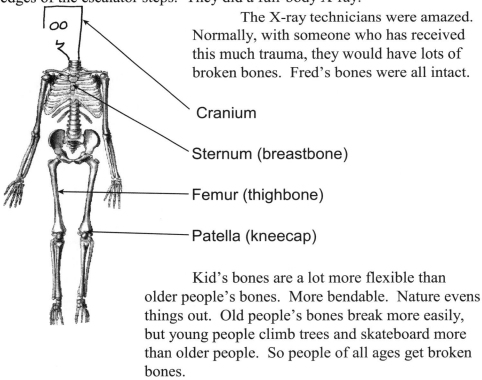

The X-ray technicians were amazed. Normally, with someone who has received this much trauma, they would have lots of broken bones. Fred's bones were all intact.

Cranium

Sternum (breastbone)

Femur (thighbone)

Patella (kneecap)

Kid's bones are a lot more flexible than older people's bones. More bendable. Nature evens things out. Old people's bones break more easily, but young people climb trees and skateboard more than older people. So people of all ages get broken bones.

Chapter Thirty-seven   Bones

Fred's nose did not show up on the X-ray because noses are made of **cartilage** and not bone. Cartilage stays flexible. A pinna (external ear) is also made of cartilage. If it were made of bone, then you might shred your pillow at night.

According to one listing, there are two-thirds of a dozen places in the body where cartilage can be found. Some biology books like to make big long lists for you to memorize. We'll skip the other six areas of the body. The nose and ears are the fun places to talk about.

Question: Who has more ribs: a man or a woman?
Answer: A woman has two dozen ribs. A man has only 24 ribs.

Question: How many bones are in the body?
Answer: That's a trick question. When you are born, you have about 350 bones. Adults have 206 bones.

Wait! Hold it! I, your reader, object. I've never seen kids in elementary school going around losing bones.

I can do the math: 350 − 206 = 144 bones lost. I can just picture my kindergarten teacher going around picking up bones that her students have lost.

The 350 little rubbery bones of newborns gradually harden and some of them fuse together as the child grows up.

Your Turn to Play

1. There are two-thirds of a dozen places in the body where cartilage can be found. Do the arithmetic. What is two-thirds of twelve?
2. Do the arithmetic: two dozen minus 24 = ?
3. Solipsists can be found in all walks of life. They believe that they are the only thing that exists. Everything else is just a show that they are seeing and hearing. Not only are they the center of the universe—they *are* the universe!

   Your pain, of course, means nothing to a solipsist. You tell them that you just lost your job, your house is being foreclosed, you have a terminal disease, and you have a blinding toothache. They may respond with something like, "That's interesting."

Your question: What argument could you present to a solipsist to convince him that he is not in touch with reality?

4. If something is a dime a dozen, how much does an individual item cost?

5. Solve $44 = 18x + 8$

........COMPLETE SOLUTIONS........

1. $\frac{2}{3} \times \frac{12}{1} = \frac{2}{\cancel{3}} \times \frac{\cancel{12}^{4}}{1} = 8$ places in the body where cartilage can be found.

2. Men and women have the same number of ribs.

3. Lots of luck. I can't think of any convincing argument. Tweaking his nose (cartilage) won't work. Everyone has physical pain in their lives. The solipsist just thinks it part of the show that he is watching.

These mind games are a dime a dozen in philosophy. Joe tells Darlene that he had a dream about a butterfly. Darlene asked Joe, "How do you know that you are not a butterfly having a dream about Joe?" Such questions are unanswerable.

4. $10 \div 12 = \frac{10}{12} = \frac{5}{6} = 83\frac{1}{3}\% = 0.8333333... \approx 0.83$ cents

($\approx$ means *approximately equal to*)

5. original equation $44 = 18x + 8$
   subtract 8 from both sides $36 = 18x$
   divide both sides by 18 $2 = x$

From the footnote of three pages ago: *Why not give food or drink to someone with a possible bone fracture?*

Doctors generally treat a broken arm or a broken femur in two steps. First, they reposition the broken ends of the bone together. In medical language, this realignment is called **reduction**. Second, they put on a cast or a splint so that the bone parts will stay aligned during the healing process.

Reduction is often done under general anesthetic.

General anesthetic causes nausea or vomiting about 25–30% of the time.

Inhaling vomit can cause pneumonia. Therefore, don't eat on the way to the hospital. You can't throw up what you haven't eaten.

Chapter Thirty-eight
Better Bones

Bones are alive. They aren't just sticks of calcium that prevent you from looking like a bowl of jelly. Bones consist of several different kinds of cells nestled in a protein and calcium superstructure. All the parts of a bone are constantly being broken down and rebuilt. The atoms in your bones right now are not exactly the same ones that were in your bones yesterday.

For about the first quarter of a century of your life, your bones are getting larger and stronger. A National Institute of Health panel recommends that adolescents get 1200–1500 mg of calcium each day.

 *Teens = 1200-1500 mg Ca/day*

What you do in your second decade of life can have a profound influence on the rest of your life.

Your bone health is one good example.

You hit your maximum bone mass at around age 25. The deposits you have made in your "bone bank" have exceeded your withdrawals during those first 25 years.

As an adult, your body can only absorb around 15% of ingested calcium.* Your days of making easy deposits in the bone bank are over.

Imagine what it would be like *if you lost a third of your bone mass*. You really wouldn't want to go skateboarding or ice skating or skiing. You wouldn't want to slip getting out of the bathtub. Your bones would break very easily.

---

* Kids can absorb up to 75% of the calcium they eat.

Chapter Thirty-eight   Better Bones

If you are an average woman, you will lose somewhere between one-third and one-half of the weight of your bones over your lifetime.

Severe loss of bone mass is called **osteoporosis**.* This loss is most severe in the hips, spine and wrists. If you have osteoporosis and you fall, it will be no surprise at the hospital if you have broken your hips, your spine and/or your wrists.

How serious is that? In 2007, the *Journal of the American Academy of Orthopaedic Surgeons* made it real clear:

> Women are more likely to die
> from fractures caused by osteoporosis
> than will die of breast cancer.

How can a teen get 1200–1500 milligrams of calcium in a daily diet? Any of the following will put you in the 1200–1500 mg. range:

☆ 4 cups of milk
☆ 3 cups of plain, nonfat yogurt
☆ 4 cups of low-fat with fruit yogurt
☆ 7 oz. of low-fat mozzarella cheese (pizza!)
☆ 8 cups of cooked broccoli
☆ 9 cups of dandelion greens
☆ 13 cups of cooked kale
☆ 5 cups of cooked spinach
☆ 21 oz. of salmon
☆ 21 oz. of sardines including the bones
☆ 4 lbs. of tofu

These are the major contributors of calcium.

---

* OS-tea-oh-purr-ROW-sis   The prefix *osteo* means "bone." So osteology is the study of skeletons. Osteomyelitis is an inflamation of bones. If there were such a word as osteoostrich, I guess it would have something to do with the bones of an ostrich.

femur?

Chapter Thirty-eight   Better Bones

Study that list. Getting enough calcium without dairy products can be a real pain in the . . . diet.

Your Turn to Play

1. Multiple-choice question. Smoking reduces the amount of oxygen reaching body tissues. A smoker's broken bones require how much longer to heal than non-smoker's bones?

    A) up to 30% longer

    B) up to 30% longer

    C) up to 30% longer

    D) up to 30% longer

2. Since eating 13 cups of cooked kale every day is not too high on my list of favorite things to eat at the dinner table (litotes), my eyes drift to the line: low-fat mozzarella cheese (pizza!). There are two major drawbacks to obtaining your calcium using low-fat mozzarella on pizza.

    Major Drawback #1: Extreme happiness may result. No one likes to be around people who are *that* happy.

    Major Drawback #2: According to the nutrition label on the package of low-fat mozzarella that I have in my fridge, there are 80 calories in every ounce, and 50 of those calories are fat calories.

    What percent of the calories are fat?

3. If you put 3 oz. of mozzarella on an 8" pizza, how many pizzas would you ~~have to~~ get to eat in order to ingest 7 oz. of mozzarella? (Use a conversion factor.)

4. Suppose the Love of Pizza is a dominant gene. Suppose one of my grandparents liked pizza and the other three didn't like pizza. Is it possible that all of their grandchildren would like pizza?

5. Suppose the Love of Pizza is a dominant gene. Suppose one of my grandparents liked pizza and the other three didn't like pizza. Is it possible that none of their grandchildren would like pizza?

. . . . **COMPLETE SOLUTIONS** . . . .

on Next Page

1. Answer B is correct. If you smoke, you can figure you'll be wearing that cast up to 30% longer. (You will also get full credit if you answered A, C, or D.)

2. 50 is what percent of 80? You divide the number closest to the *of* into the other number. $\frac{50}{80} = \frac{5}{8} = 62.5\%$ (This is one of the conversions you memorized.)

3. $\frac{7 \text{ oz.}}{1} \times \frac{\text{one pizza}}{3 \text{ oz.}} = \frac{7}{3}$ pizzas $= 2\frac{1}{3}$ pizzas

4. Let's let P stand for the pizza-loving gene.
Let p stand for the recessive gene of not liking pizza.

If Grandma Katherine was the one who liked pizza, she would either possess PP or Pp. Her husband, since he didn't like pizza, would have to be pp.

Grandma Katherine's son would then either be Pp or pp, depending on whether Grandma Katherine passed on to him her P or her p gene.

If Grandma Katherine's son was Pp, then he could conceivably pass on the P gene to each of his kids.

5. In the previous answer, we established that Grandma Katherine's son was either Pp or pp. My other three grandparents didn't like pizza, so they all had to be pp.

With Grandma Katherine's son being Pp (or pp), and his wife being pp, it is possible that all their children would be pp and not like pizza.

Would a diagram help?

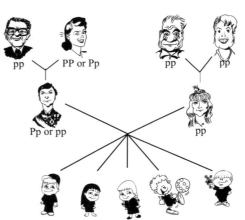

All the grandkids are either Pp or pp.
It's possible for them all to be Pp or all pp.
It is most likely that some will be Pp and some pp.

Chapter Thirty-nine
Integumentary System

Fred's skeletal system was intact. He hadn't turned 206 bones into 207 bones by breaking one of them. But his integument was a mess. It was covered with marks that looked like iiiiiiiiiiii.

Before you, my reader, interrupt me, let me define. . . .

Hey! Before you go any farther—or is it further? Let me ask you what "integument" means.

Intermission

There are three levels of English: Informal English, General English, and Formal English.

Informal English is usually spoken. When used in public, it indicates some kind of close relationship between the parties. Salesmen may use Informal English a lot to establish rapport (pronounced ra-POUR). Words like *in-laws*, *doohickey*, and *ain't* are used.

For individuals using Informal English, the distinction between *farther* and *further* is the furthest thing from their minds.

In General English, you use the rule: Never use *farther* unless you are talking about physical distance. That means that you can always use *further*.

Formal English is used in scholarly writing or in speeches to educated audiences. In Formal English, *farther* is used for physical distances, and *further* is used for figurative distance.

"He spoke further on the topic."

Chapter Thirty-nine Integumentary System

Integument is the outer covering of something. Fred's integument is his skin, hair, and nails. [in-TEG-yeh-ment, where TEG rhymes with *keg*.]

A secondary definition of *integument* is an enclosure. If you have just rented a new little apartment, you might send out to your friends, "Come celebrate my new integument." Of course, they might think you were announcing that you just got a facelift.

They wheeled Fred from the X-ray department to a hospital bed. As Fred lay there counting the ▦▦▦ marks on his integument, Sally knocked on the door and came bouncing in. She brought a bouquet.

"Hello, Mr. Man," Sally said. "I heard you fell down the 'skater."

Fred thanked Sally for bringing the flowers. She continued to hold on to them. She didn't know that you were supposed to *give* the flowers. She thought you were just supposed to show them to the sick person and then take them back to the florist.

She looked at one of the purple ▦▦▦ marks and poked it. "Did the 'skater make those marks?" she asked.

Fred moaned.

Sally had stimulated some of the sensory neurons in Fred's dermis.

Translation:

Neurons are nerve cells.

Sensory neurons are nerve cells that detect things like heat, cold, pressure, or pain.

Dermis is the layer of skin right underneath the epidermis.

The epidermis is the outer layer of skin.

The door opened again. Jimmy looked in and yelled down the hall, "Mom. She's in here."

Sally gave a wink and headed out the door.

A second later, Fred could hear Sally's mother, "Where were you? I thought you were at the mall. And where did you get those flowers? Did you take it from that patient? Here Jimmy. Take those flowers and put them back in that room."

Wordlessly, Jimmy entered Fred's room and tossed the bouquet on an empty chair. As he left, the flowers fell on the floor.

Sometimes, Fred thought, life feels like flowers on the floor. (simile)

Chapter Thirty-nine   Integumentary System

Your Turn to Play

1. When I was about ten years old, I asked my best friend, "What's worse—being blind or being deaf?" He answered, "What?" and giggled.

    There are various sensory neurons in your dermis. Some detect warmth. Some cold. Some pressure. And some pain. What is the most valuable of those neurons?

2. Name the largest organ in your body. In adults it weighs about nine pounds. It is not your heart, or your liver, or your brain.

3. The epidermis is about 0.1 millimeters thick. Express that in inches. (Conversion factor: 1 mm = 0.039 inches)

4. The epidermis isn't always about 0.1 millimeters thick. Some places it's a lot thicker. Where?

. . . . COMPLETE SOLUTIONS . . . .

1. It's *nice* to feel a warm sun on your back in the springtime. In the summer, it's *nice* to feel a cool breeze from the air conditioner. It's *nice* to feel the touch from someone you love. But it is almost *absolutely essential* to feel pain if you want to stay alive for long.

I, your reader, know of a lot of times where I wish I had no pain receptors. With no pain receptors, I could say to my mother, "I really wouldn't mind an extra spanking today."

    But when you lean on a hot stove, those neurons in your hand that detect pain keep you from getting your hand cooked.

    When you break your femur, your pain nerves scream, "Don't move me! Don't touch me! Don't even look at me! In fact, don't even think about me!" Those pain nerves are announcing the danger that the sharp ends of your broken thigh bone could do. Without those nerves, you could continue for about three steps . . . and probably bleed to death.

2. It is the part of your integumentary system that is called your skin.

3. $\dfrac{0.1 \text{ mm}}{1} \times \dfrac{0.039 \text{ inches}}{1 \text{ mm}} = 0.0039$ inches

4. Go barefoot a lot, and the bottoms of your feet can become almost like shoe leather. Calloused hands are nature's gloves—a perfect fit.

Chapter Forty
Epidermis

Fred knew that the ||||||||| marks on his epidermis would go away. The cells on the top of the epidermis are flat and dead. They are constantly wearing away. Underneath, the skin cells are constantly dividing like crazy.

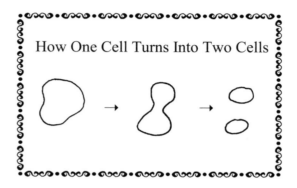

This is called **mitotic division**. The two daughter cells look exactly like the parent cell. If the parent's eyelash genes were LS (where L is the dominant gene for long eyelashes, and S is the recessive gene for short eyelashes), then each daughter cell will have the LS pair of genes.

In the olden days, you might say that each daughter cell was a carbon copy of the parent cell, but nowadays, how many people own a sheet of carbon paper? In computer-talk, we might say that mitotic division is Cut followed by two Pastes.

So the epidermal* cells are dividing like crazy at the base of the epidermis. They push their way upwards towards the surface. It reminds me a little of popcorn being popped in a pot: they are all shoving their way to the top.

Their trip to the top takes about a month. Recall from the previous *Your Turn to Play*, we are talking about a trip of 0.0039 inches. When they

---

★ *Epidermal* is the adjective form of epidermis. Adjectives are words that describe.

Chapter Forty   Epidermis

have made it to the top, they are all tuckered out, flattened, and dead. (This is in contrast to popcorn, which is all lively and fluffy.)

But you want dead, flattened out epidermal cells on the surface of your epidermis.

I do?

Yes. For several reasons. First, as these dead cells flake off, they help erase the ▨▨▨▨ marks. Most of the time you don't notice the cells flaking off. We have a special name for the skin cells that we notice flaking off our heads: *dandruff*.

Second, we love those dead, flattened out epidermal cells because they help keep out the bad bacteria, viruses, and toxins. Little kids sometimes think that the purpose of skin is to hold their innards in, a little like the skin of a hot dog holds the meat in. When have you ever heard a little kid say . . .

> The unbroken epidermis provides a significant barrier to the incursion of deleterious agents into one's body.

Hold it! I, your reader, have to object to your poor English in the previous question. I don't mind your using big words like "deleterious," but you wrote a question and didn't end the sentence with a question mark. Why not?

I couldn't think of where to put it.

You got to put it somewhere.

Okay.

I'm glad this is not some stuffy old book where the reader can't ask questions. I have a second question.

As Don Vito Corleone once said, "When have I ever refused an accommodation?"

You wrote that it takes about a month for the newly created epidermal cells to travel from the bottom to the top of my epidermis, which

Chapter Forty   Epidermis

is a trip of 0.0039 inches. That's easier for me to think about than 0.1 millimeters, but 0.0039 inches is hard to visualize.

Let's stack a thousand epidermal layers together. How tall would that be?

0.0039 × 1000 = 3.9 inches

A thousand sheets of paper is two reams of paper.

Hey! Two reams of paper are almost four inches tall.
That means my skin is the thickness of a sheet of paper.

No. It means that your epidermis is as thick as a sheet of paper. Your dermis is much thicker.

A second function of your epidermis—besides keeping the bad bacteria, viruses and toxins out—is to protect the layers and tissues underneath from friction and from blows. In the animal kingdom, human skin is a very poor protector. Look at Fred. Just falling down an escalator has covered his body with ▭▭▭▭ marks.

In contrast, a rhinoceros has really great skin. Rifle bullets will sometimes fail to penetrate rhino skin. You can guess how many ▭▭▭▭ marks a rhino would have if he fell down an escalator.

The drawback in having rhino skin is that you have to use a lot of hand lotion if you want your skin to be soft and silky.

Chapter Forty   Epidermis

Let's suppose we had a big sheet of epidermis lying around that nobody was using. Let's thicken it up by folding it in half. It would now be 0.0039 × 2 = 0.0078 inches thick. That's not quite thick enough to deflect a rifle bullet. (litotes)

Okay. Let's take a sheet of epidermis and fold it in half fifty times. That means fold it in half. Then fold it in half again. Then fold it in half again. Then fold it in half again. Then fold it in half again. Then fold it in half again. Then fold it in half again. Then fold it in half again. Then fold it in half again. Then fold it in half again. Then fold it in half again. Then fold it in half again. Then fold it in half again. Then fold it in half again. Then fold it in half again. Then fold it in half again. Then fold it in half again. Then fold it in half again. **Fifty times.**

Can you imagine how thick that would be?

We start with 0.0039 inches. Then double it. Then double it again. Fifty times.

Fun Quiz Time! Anyone who hasn't stayed up too late at night or hasn't ingested too much sugar should be able to make reasonable estimates of things as simple as the effects of doubling something several times. Here is your chance to prove that you are not an over-sugared owl.

If you double a 0.0039 inch epidermis fifty times, which of the following is true?

        A) An arrow could penetrate all the layers.

        B) It would take a rifle bullet to get through all the layers.

        C) It would take an H-bomb to get through all the layers.

        D) An H-bomb would hardly make a dent.

What we are looking for is $0.0039" \times 2 \times 2 \times 2 \times 2 \times 2 \ldots \times 2$, which is $0.0039" \times 2^{50}$.

Algebra fact: $2^3 \times 2^4$ is equal to 2^7. (Not 2^{12}.) It's easy to prove. Proof: $2^3 \times 2^4 = (2 \times 2 \times 2) \times (2 \times 2 \times 2 \times 2)$
$= 2 \times 2 \times 2 \ \times \ 2 \times 2 \times 2 \times 2$
$= 2^7$.

In general, $x^a x^b = x^{a+b}$. (You'll notice that the minute I start using x, I can no longer use ×.)

So $0.0039" \times 2^{50} = 0.0039" \times 2^{10} \times 2^{10} \times 2^{10} \times 2^{10} \times 2^{10}$.

Chapter Forty   Epidermis

I chose 2^{10} because 2^{10} is really nice. It's equal to 1024. (I have used that fact so many times over the years, that it has memorized itself.)

$$\begin{aligned}\text{So } 0.0039" \times 2^{50} &= 0.0039" \times 2^{10} \times 2^{10} \times 2^{10} \times 2^{10} \times 2^{10} \\ &= 0.0039" \times 1024 \times 1024 \times 1024 \times 1024 \times 1024 \\ &\approx 0.0039" \times 1000 \times 1000 \times 1000 \times 1000 \times 1000 \\ &= 0.0039" \times 10^3 \times 10^3 \times 10^3 \times 10^3 \times 10^3\end{aligned}$$

which (by the algebra rule on the previous page) is $0.0039" \times 10^{15}$.

That's a lot of equations to look at. Please don't just zip your eyes over the five lines of the previous paragraph. Seeing what happened on every line is important.✶

Your Turn to Play

1. $0.0039" \times 10^{15}$ is the same as multiplying 0.0039 by 10, fifteen times. It moves the decimal point 15 places to the right.
   Do it. Write $0.0039" \times 10^{15}$ without the "$\times 10^{15}$."
2. Write out 0.0039×10^{15} in words.
3. Doubling a 0.0039 inch piece of epidermis fifty times will make a skin sandwich that is 0.0039×10^{15} inches thick. Doing the arithmetic to change this into miles is too much.
   Just write 0.0039×10^{15} times the conversion fractions, but don't work it out. (12" = 1' and 5280' = 1 mile.)

✶ You read math m u c h m o r e s l o w l y than you read English novels. One course at the university I went to was entitled, "British Novels of the 18th Century." In the bookstore, there were twelve novels you had to buy for that course. (I didn't take that course.) And that was just one course.

English majors have to read zillions of pages and get short fingers typing long term papers.

Math majors read a tiny fraction of the pages that English majors read. No long term papers. The only difference is that math majors have to pay attention to the few lines that they have to read.

Chapter Forty   Epidermis

> .... **COMPLETE SOLUTIONS** ....
>
> 1. $0.0039" \times 10^{15} = 3{,}900{,}000{,}000{,}000$ inches.
>
> 2. Three trillion, nine hundred billion.
>
> 3. $\dfrac{3{,}900{,}000{,}000{,}000 \text{ inches}}{1} \times \dfrac{1 \text{ foot}}{12 \text{ inches}} \times \dfrac{1 \text{ mile}}{5280 \text{ feet}}$

We are interested in *roughly* how many miles thick fifty doubles of skin will be. We can take the answer to question 3 and approximate the answer **WithoUt NeeDiNG a CaLCULatoR**.

We start with $\dfrac{3{,}900{,}000{,}000{,}000 \text{ inches}}{1} \times \dfrac{1 \text{ foot}}{12 \text{ inches}} \times \dfrac{1 \text{ mile}}{5280 \text{ feet}}$

Twelve divides into 39 about three times:

$$\dfrac{\overset{3}{\cancel{3{,}900{,}000{,}000{,}000}} \cancel{\text{inches}}}{1} \times \dfrac{1 \text{ foot}}{\cancel{12} \cancel{\text{inches}}} \times \dfrac{1 \text{ mile}}{5280 \text{ feet}}$$

$$\approx \dfrac{300{,}000{,}000{,}000 \text{ feet}}{1} \times \dfrac{1 \text{ mile}}{5280 \text{ feet}}$$

Approximate 5280 by 5000:

$$\approx \dfrac{300{,}000{,}000{,}000 \text{ feet}}{1} \times \dfrac{1 \text{ mile}}{5000 \text{ feet}}$$

$$= \dfrac{\overset{6}{\cancel{300{,}000{,}000{,}000}} \text{ feet}}{1} \times \dfrac{1 \text{ mile}}{\cancel{5000} \text{ feet}}$$

$= 60{,}000{,}000$ miles

This is about two-thirds of the way to the sun. Please remember to start out with a fairly large piece of epidermis for this experiment.

H-bombs wouldn't make a dent in this much skin.

Chapter Forty-one
Dermis

Fred hopped out of bed, took the flowers and put them on the night stand next to his bed. Sometimes, Fred thought, life feels like flowers given to you by a friend.

The nurse came into the room. "The X-rays came back just fine. You didn't break any bones. You didn't even break your skin. You just have some bruises, and you don't need to be in the hospital to have them heal."

Fred knew what this meant. He put on his shoes, thanked the nurse, and raced down the hallway. Normally, he would have taken the stairs "for the exercise," but today, he elected to take the elevator.

Out into the sunshine, Fred started to sing a song. He made up the words as he sang:

♪   My bones are two-oh-six.
♪   My skin unbroken is.
      I'll dance and do some kicks,
      And now I'm back in biz.   (*biz* is short for business)

The melody Fred invented matched the quality of his poetry. Fred made up for these defects by singing loudly.

A driver stopped his car in the middle of the street and called out to Fred, "Son, are you in pain?"

Fred thought the driver was referring to his purple iiiiiiiiiiiii marks and answered, "Yes, they hurt, but it's okay. Thanks for asking."

The driver thought to himself, The kid must be nuts, and drove on.

Those purple marks extended down into Fred's dermis, which is the layer of skin below the epidermis. The dermis is much more exciting than the epidermis. Lots more things are happening "down there" 0.0039 inches below the surface of the skin.

Chapter Forty-one   Dermis

The epidermis has no blood vessels. It is the blood vessels in the dermal* layer that supply the oxygen and food to the underside of the epidermis (noun) where the epidermal (adjective) cells are making new skin.

If the dermis didn't share its oxygen and food with the epidermis, then mitotic division in the epidermal cells would stop.

If epidermal mitotic division stopped, there would be no new skin cells.

If there were no new skin cells, the 0.0039" epidermal layer would soon wear away.

If that protective, waterproof, germ-proof, layer were missing, the rich, red, bloody dermis would be exposed to all the bad bacteria and viruses that are out there.

The bacteria and viruses would yell, "Buffet time!" (This is what happens on a minor scale when you get a cut that gets infected.)

Your body would no longer function properly, since you would be dead.

Are you glad your dermis shares its oxygen and food?

If you were stitching a sampler, it might look like this:

It's nice to share.

The dermis makes three things that poke through the epidermis. Quiz time again. How many of those three can you name? I'm going to leave the rest of this page blank so you can think.

[left blank intentionally]

---

* *Dermal* is the adjective form of dermis. Or we could say that *dermis* is the noun that corresponds to the adjective *dermal*. Nouns are words that are persons, places, or things.

217

Chapter Forty-one   Dermis

Did you stop and think? Or did you just turn the page and continue to read?

Thinking is hard work. Many people avoid it as much as possible. Some will, for example, read a question in the *Your Turn to Play* and then just skip down to read the answer.

There is an alternative to a life filled with thinking. It is a life filled with just physical work. You get to take your choice.

Most people can think of two of the three things that the dermis makes that poke through the skin. In a minute, you will be able to name all three.

First is **hair**. We don't have much hair compared with many mammals.* Hair is great for keeping many mammals warm. Did you ever see a dog shivering?

Attached to each hair follicle in the dermis of mammals is a small muscle. When the animal gets cold, the muscle makes the hair "stand on end." The coat gets fluffier and holds in the warmth better.

We have those muscles in our dermis. When we get cold, all we get are goose bumps. Cattle and bears and dogs play in the snow without putting on overcoats. Naked, we do pretty well between 74° and 74.001°.

Second are **sweat glands**. Down in the dermis, sweat glands produce . . . sweat. It's 99% water with a little salt mixed in. There are 2,500,000 sweat glands, and they kick into high gear when you get warm. The sweat passes through a duct (a tube) in the epidermis. When it evaporates, it cools you off.

When you become an adult, you get a second set of sweat glands. At age 5½, Fred doesn't have this second set yet. This second set of sweat glands appear in the places where adults have hair that kids don't. These glands produce a sweat that contains more than just water and salt. It also

---

* Mammals are the part of the animal kingdom that: (1) have hair; (2) breast-feed their young; (3) are warmblooded; (4) have complex teeth—incisors for cutting, canine teeth (eye teeth) for tearing, molars for grinding; and (5) have a highly developed brain (cerebrum).

You are a mammal.

contains proteins. Delicious proteins—at least the skin bacteria think so. Initially, this sweat is odorless, but after the bacteria feast on it for a day or so, the bacteria poop starts to smell.

In the olden days, the only way adults could avoid knocking out others with their body odor was to bathe. About a century ago, antiperspirants and deodorants appeared in the stores. Antiperspirants stop somewhere between 15% and 50% of sweating. Deodorants either help kill the bacteria or just perfume the area. Both can cause allergies or skin irritation in some individuals.

The third thing that the dermis makes that pokes through the skin is **oil glands**. The oil glands produce . . . oil. Just like the sweat glands, they have tiny ducts which transport the goodies through the epidermis to the surface of the skin.

If you are going to be a biology major, you have to call the oil glands the **sebaceous glands**. The hardest part of being a biology major is memorizing all the terminology. I, personally, prefer "oil glands" to "sebaceous glands" because: (1) I'm not a biology major; (2) *oil* is shorter than *sebaceous*; (3) I don't know how to pronounce *sebaceous*; and (4) oil glands are much more descriptive to me than sebaceous glands.

What is good about ~~sebaceous glands~~ oil glands is that the oil makes your hair all nice and shiny, and it makes your skin smooth and pliable.

What is rotten about oil glands is that those bad bacteria also like to dine on oil (in addition to enjoying the protein provided by the adult sweat glands). When they infect the oil duct glands, the skin gets inflamed. It's called acne.

The dermis does one thing that doesn't involve a poke through the epidermis. Suppose the weather looks like this and you step outside wearing just your pajamas. There is a small chance you may freeze to death. Your body doesn't like that. In order to avoid losing heat, the blood vessels in your dermis contract so heat is not lost to the outside air. You may turn a lovely blue.

Chapter Forty-one   Dermis

In contrast, suppose it's a little toasty outside. Again, your body doesn't like the idea of dying. It rushes tons of blood to your dermis so that it can be cooled off.

"Tons of blood" is a bit of an exaggeration. In English that's called **hyperbole**. [high-PERB-bow-lee] A ton is 2000 pounds, and not too many individuals have 2000 pounds of blood in their bodies. (litotes)

What is surprising is that when you are feeling really warm, the blood vessels in your dermis dilate so much that 50% of your whole body's blood supply will be in your skin. That leaves very little blood for your brain or your muscles. This paragraph is your 𝕺𝖋𝖋𝖎𝖈𝖎𝖆𝖑 𝕰𝖝𝖈𝖚𝖘𝖊 for sitting around and watching mindless television when the weather gets really hot.

Your Turn to Play

Given information: Let L be the dominant gene for long eyelashes. Let S be the recessive gene for short eyelashes.

Suppose the father had long eyelashes. Suppose he and his wife had six children, and one of the six children had short eyelashes.

1. What is the genotype of the father? In other words, is he LL, LS, or SS?
2. What is the genotype of the mother?
3. Could the mother have long eyelashes?
4. One of the kids is two meters tall. Is that a tall child?
5. The extremes are often a bad place to be in life. If you bathed once every four months, no amount of antiperspirant or deodorant would work, and you would probably be overwhelmed by the bad bacteria. The score would be Bacteria: 1 and You: 0. In this case, the most significant aspect of your dermis would be the adult sweat glands.

On the other hand, if you bathed daily, the score would be Bacteria: 0 and You: 1. In this case, what would be the most significant part or parts of your dermis?

Chapter Forty-one   Dermis

········COMPLETE SOLUTIONS········

1. Since one child has short eyelashes, that child's genotype must be SS. (If it were LS or LL, then the child would have long eyelashes, since L is dominant over S.) Therefore, the child received an S from each parent. So the father must have at least one S in his genotype. We are told that he has long eyelashes, so he must also have at least one L. Dad is LS.

2. Since one child has the genotype SS, that child received an S from each parent. So the mother has at least one S. So she is either LS or SS.

3. Since the mother is either LS or SS, she might have either long or short eyelashes.

4. A meter is a little longer than a yard. The child is more than two yards tall.

   Actually, a meter is a little longer than 39 inches. So the child is a little taller than 78", which is 6'6". I'd call that tall.

5. Your oil glands would probably be working full time if you bathed daily.

   If you bathed five times a day, your oil glands couldn't produce oil fast enough. The second that any little bit of oil made its way from the oil glands through the ducts up to the surface, it would get soaped away. Your hair wouldn't be nice and shiny, and your skin would be like unbuttered toast.

   That's why frequent bathers tend to buy a lot of hair conditioners, body oils, and lotions.

   Just think for a moment about those cowboys who would ride for twenty days across the prairie herding cattle. Try selling them some hair conditioner! Their hair is already plenty shiny.

Cowboy about to sneeze. Notice how he covers his mouth so he won't infect his horse.

Chapter Forty-two
Genes

Fred started singing the second verse of the song he had made up. "♪Unbroken bones ♫ are fine with me. . . ." A piece of paper blown by the wind interrupted his singing. He read the first line, "Get Rich!! C.C. Coalback's Easy Road to Riches" and knew this advertisement was phony. He removed the paper from his nose and tossed it into a nearby garbage can.

He giggled as he thought to himself *If I ever am in a choir, I wouldn't have to hold my sheet music like everyone else.*

Having paper get stuck on his nose was so common for Fred, that he hardly gave it a second thought. What happened next was totally new for Fred.

Fred had accidentally speared a fruit fly. It wasn't a regular fruit fly. This one had purple eyes. (If you, my reader, happen to own a purple crayon or pen, and *if you are the one who purchased this book*, then you can help the story along by coloring the fly's eyes purple. We have kept the price of this book low by not filling it with colored illustrations.)

The fruit fly *Drosophila* does not normally have purple eyes. This fly was a **mutant** (= result of a mutation).

In a normal population of fruit flies, there are no genes that produce purple eyes. But every once in a while, the eye color gene experiences a mutation. A **mutation** is a change in a gene that gets inherited by future generations.

∗ I'm not sure whether it is legal to attach a footnote to someone's head, but I need to mention that fruit flies are generally much smaller than the one pictured here. They are smaller than houseflies.

For those of you who eat boxed cereal, you will instantly recognize my statement: "Enlarged to show texture."

Chapter Forty-two   Genes

Biologists love to play with *Drosophila*. (Another reason I am a mathematician and not a biologist. Playing with flies has never been "my thing.") *Drosophila* are cheap. They breed easily. The generations are short. And they are easy to feed.

If they capture some of these purple-eyed mutants, they can breed whole populations of purple-eyed fruit flies. (The one that Fred impaled on his nose wouldn't be good for breeding. It's dead.)

These mutants don't do as well out there in the real world as regular fruit flies. If the biologist who has bred thousands of purple-eyed fruit flies decides to let them go, then: (1) the neighbors may complain, and (2) they will occupy less and less of a percentage of the whole *Drosophila* population as the days go by. They will die out.

Every once in a while in the regular fruit fly population a mutation will occur, and suddenly there is a purple-eyed fly in the crowd. Mutations are a regular occurrence in animals, plants, and viruses. Happily, mutations happen rarely. When people talk about getting some "R&R," they usually are thinking of rest and recreation.* With regard to mutations, RbutR could mean regularly but rarely.

Each cell in your body has the same set of genes as every other cell in your body. A tiny exception to that rule occurs in a funny kind of cell division called meiosis—each resulting cell has only half the set of genes. Those cells are not good for anything . . . except making babies.

But every other part of the body—brains, liver, skin—have been formed by mitotic division. This process—**mitosis**—preserves the whole set of genes. If, for example, your cells contained one dominant eyelash gene L and one recessive eyelash gene S, then after mitosis, each daughter cell would have LS.

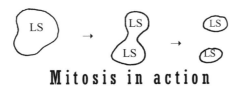

Mitosis in action

---

* R&R can also mean rest and recuperation, or it can mean rest and relaxation, or it can mean rock and roll.

Chapter Forty-two — Genes

Wait a minute! That means I got eyelash genes in my brain cells!

Yup. Live with it.

But that's stupid. Why would I want my hair color genes to be in my tongue?

Until recently, the appropriate answer to your question would be that your wants have little to do with biological reality. In mathematics, you may want 2 + 2 to equal 5 on Thursdays, but there is little chance of your desires becoming reality.

In contrast, in biology, things aren't as fixed. Maybe tomorrow, you will be able to eliminate your hair color genes from your tongue cells.

You tell me that my dermis cells and my eyeballs both have the same set of genes. How come I don't have hairy eyeballs?

I don't know.

But I have a "fake" answer. We know that certain genes can affect the expression of other genes. Certain genes can tell the hair-growing genes to "Shut up." In the eyeball, those controlling genes are active, and you don't grow hair on your eyeballs.

I can see why you called that a "fake" answer. If all cells have the same set of genes, then the genes that tell the hair-growing genes to "Shut up" are also present in the dermis. Why aren't they active in the dermis?

Like I said, I don't know.

Let me go back a moment to when I described mutations as RbutR—regularly but rarely—and talk about those two words.

To say that mutations are a **regular** occurrence in animals, plants and viruses does *not* mean that they occur like clockwork. You can't time mutations like cars that come off an assembly line every 26 seconds. That definition of **regular** is the fourth definition of the word in my dictionary.

When we say that mutations are **regular**, we mean that they are unsurprising. They are a normal part of reality. They are like babies needing to have their diapers changed. This is the first definition of **regular** in my dictionary, except that they left out the part about diapers.

You don't find biologists running to the newspaper office shouting, "I found a mutation!" Mutations are expected as long as there is life.

Chapter Forty-two Genes

On the other hand, mutations are a **rare** occurrence in animals, plants, and viruses. Take a look at any gene. For example, there is the gene for eyelash length, or the gene that produces some particular protein in the body. Very roughly speaking, if you look at that gene in the parents and then look at that gene in their child, the chances that you will see a mutation is somewhere between one in 100,000 and one in 1,000,000.

Some genes mutate much more frequently than others.

The effect of mutations? Many mutations have no discernable effect. Change one gene, and nothing seems to change.

Before mutation

After mutation

Mutations are **random**—besides being **regular** and **rare**.

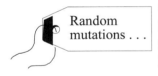 are like taking the remote control while your sister is watching television and pushing a button at random. Sometimes nothing happens. Or, if you happen to hit the power button, the results can be fatal to her television viewing (and to you).

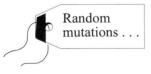 are like taking a piece of art and moving some line or point. If you were to make a random change in this piece of fine art . . .

would it improve it?

fine art

225

Chapter Forty-two   Genes

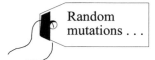 **Random mutations . . .** are like taking a sentence and making a random change in in it. (The error in the previous sentence is deliberate.)

Your Turn to Play

1. **Point mutations** are mutations that occur at one particular point inside a gene. If genes were words, then a point mutation would be the change in a single letter of a w<u>u</u>rd.

Color blindness,

cystic fibrosis,

sickle-cell disease, and

Tay-Sachs disease are each the result of a point mutation.

Some mutations involve the deletion of one or more genes. The cri du chat syndrome is an example. Children born missing a part of the genes on chromosome 5 (I'll define *chromosome* in the next chapter) get this disease. Infants with cri du chat have a characteristic cry that sounds like a meowing kitten.* Cri du chat occurs in one birth in somewhere between 20,000 and 50,000.

Change $\frac{1}{20,000}$ to a percent.

2. Suppose there is a dominant gene, G, that produces some essential protein in the body. Suppose the recessive form of that gene (call it g) doesn't produce that essential protein. By "essential" we mean that if you don't have that protein, you would die.

Question 2a: Is it possible to have a parent with a gg genotype?

Question 2b: Is it possible to have a parent with a Gg genotype?

Question 2c: If each parent had the genotype Gg, what might we expect of their children?

Question 2d: If one parent was Gg and the other was GG, what would their children be like?

---

* Individuals with cri du chat syndrome may also have a whole host of other symptoms affecting their thinking ability, their speech, their coordination, their hearts. In French, cri du chat means cry of a cat. (I don't know how to pronounce *cri du chat*.)

Chapter Forty-two   Genes

........COMPLETE SOLUTIONS........

1. First, we need to change the fraction into a decimal. There are several ways to do that. Long division is one method. It always seems to work.

$$20000 \overline{)1.00000} \quad 0.00005 \quad \quad 0.00005 = 0.005\%$$

A second method would be to multiply the top and bottom of the fraction by 5: $\frac{1}{20,000} = \frac{5}{100,000}$ which is 0.00005 which is 0.005%.

2a. If a person had the genotype gg where g doesn't produce an essential protein, then that person would die—either before birth or shortly thereafter. They would never live long enough to become parents.

The form of a gene, such as g, that can't produce something that is necessary for life is called a **lethal allele**. [a-LEEL]

2b. If a person has the genotype Gg and g is a recessive lethal allele, then since G is dominant, it hides the bad effects of g. You can't tell by just looking whether the person is GG or Gg.

In the olden days before discussion of bad recessive genes was a frequent topic of conversation at the dinner table, mothers would tell their daughters to look carefully at the family of her potential boyfriend. "He comes from a good family" was a common expression. That was good advice. If some of his relatives had things you didn't like, they might be hidden in his genotype.

genotype = the genetic makeup of an individual. For example, they may have one long eyelash gene L and one short eyelash gene S.

phenotype = the physical expression of one's genotype. The phenotype is "what shows." Since the long eyelash gene L is dominant over the recessive short eyelash gene S, a person with a genotype of LS will have a phenotype of long eyelashes.

So in the olden days, the advice, "Check out his family," was really a shorthand way of saying, "Examine the phenotypes of his close relatives in order to get a better picture of your boyfriend's genotype."

Sometimes too much biology can be a bad thing. It's much better to write: *I love the color of your eyes in the moonlight. Your hair, your skin, your eyes are so beautiful*—instead of writing: `You have a desirable phenotype.`

2c. If each parent has the genotype Gg, then the father might contribute either G or g, and the mother might contribute either G or g. There are four equally likely possibilities:

  Father contributes G, and mother contributes G. Kid is GG.
  Father contributes G, and mother contributes g. Kid is Gg.
  Father contributes g, and mother contributes G. Kid is gG.
  Father contributes g, and mother contributes g. Kid is gg.

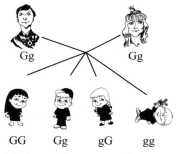

The first three kids have the same phenotype. The last kid is dead, which is a different phenotype.

2d. If the father is Gg and the mother is GG, the father's contribution is either G or g. The mother's contribution is always G.

There are only two equally likely possibilities. The kids will either be Gg or GG.

Chapter Forty-three
Six Words

Fred stood there with this fruit fly impaled on his honker. Before he could remove it, he heard a now familiar voice, "Hello, Mr. Man. That is some booger you got there." It was Sally. The blood supply in the dermis in Fred's face increased greatly in his embarrassment. (That's called blushing.)

"Would you like my hanky?" Sally proffered a pink handkerchief.

The etiquette books are strangely silent on what to do when a two-year-old offers a five-and-a-half-year-old her handkerchief in order to remove a purple-eyed *Drosophila* from one's nose. Fred was thinking Do I wipe it off and just return the handkerchief to her? Or do I retain the handkerchief and wash it before returning it? Do I just remove the fly with my hand and put it in my pocket?

As Fred just stood there, Sally reached up and wiped his nose for him.

Most two-year-olds aren't heavily concerned with etiquette. (litotes) In fact, she couldn't care less. So she stared at the contents of her hanky. (This is something that every etiquette book proscribes. You are not supposed to carefully examine your—or anyone else's—nasal productions.)

"Hey," Sally said, "This is a mutant form of *Drosophila*. I got a question. Do you happen to know which chromosome of the fly was affected, and whether there are many alleles for eye color at that locus?"

Fred was amazed. He asked, "How did you learn all those fancy words? Most two-year-olds would just talk about 'some funny-looking fly with purple eyes.'"

"My brother Jimmy—he's four years old. He was reading me all those fancy pairs of genetics words like *locus* and *site*, and

DNA and *chromosome*, and

gene and *allele*.

He told me that everyone mixes up those pairs of words. But we got those words all straightened out in the biology book he read to me. It's his favoritest biology book. It explained all those words in Chapter 43."

Chapter Forty-three Six Words

Sally's mom appeared. "Where have you been? Jimmy and I have been looking all over for you." She yanked Sally's arm, and they headed off.

Jimmy had his book with him, and Fred could hear Sally asking, "Read me that chapter again."

Jimmy began, "Fred stood there with this fruit fly impaled on his honker. . . ."

Those words sounded so familiar to Fred. He walked back to the math building and climbed the stairs to the third floor. He walked down the hallway, passed the vending machines and entered his office. The walls of his office were filled with books. In the biology section of his library were:

A Guide to Things Biological by Prof. Eldwood

Biology Reexplained by Prof. Eldwood

Looking at the Living by Prof. Eldwood

Directions for Dissection by Prof. Eldwood

Then Fred spotted the book that Jimmy had been carrying. He skipped to the bottom of the second page in Chapter 43 and began reading:

There are six words that get used a lot in biology, but which most people have only the haziest idea of what they mean. If you ask someone who is a high school graduate, "What is a gene?" they might answer, "A gene is a thingy inside of a cell."

If you ask them, "What is a chromosome?" they might answer, "It is sort of like a gene. It's a thingy inside of a cell."

If you ask them, "What is DNA?" they might say, "DNA is short for deoxyribonucleic acid. [dee-OX-ee-RYE-bow-new-CLAY-ic] It has something to do with chromosomes. It's a thingy inside of cells."

Fred skipped to the third page in Chapter 43.

Chapter Forty-three   Six Words

We will start with the big picture and work our way down to the atomic level.

Big Picture: We start with **chromosomes**. They come in pairs.* If you stick a cell from your body under a good microscope, you can see your chromosomes. They will look like a pile of rubber bands.

Before 1922, scientists were playing the game: Guess the number of pairs of chromosomes in a human cell. Some scientists counted 4 pairs. Some counted more than 25 pairs.

In 1922, Theophilus Painter announced that there are 24 pairs. Finally, the issue was settled. Human beings have 48 chromosomes. All the textbooks declared: 48 chromosomes. In every biology class: 48 chromosomes. On every biology test, the correct answer was 48 chromosomes.

In 1922 everyone knew we had 48 chromosomes.
In 1923 everyone knew we had 48 chromosomes.
In 1924 everyone knew we had 48 chromosomes.
In 1925 everyone knew we had 48 chromosomes.
In 1926 everyone knew we had 48 chromosomes.
In 1927 everyone knew we had 48 chromosomes.
In 1928 everyone knew we had 48 chromosomes.
In 1929 everyone knew we had 48 chromosomes.
In 1930 everyone knew we had 48 chromosomes.
In 1931 everyone knew we had 48 chromosomes.
In 1932 everyone knew we had 48 chromosomes.
In 1933 everyone knew we had 48 chromosomes.
In 1934 everyone knew we had 48 chromosomes.
In 1935 everyone knew we had 48 chromosomes.
In 1936 everyone knew we had 48 chromosomes.
In 1937 everyone knew we had 48 chromosomes.
In 1938 everyone knew we had 48 chromosomes.
In 1939 everyone knew we had 48 chromosomes.
In 1940 everyone knew we had 48 chromosomes.
In 1941 everyone knew we had 48 chromosomes.
In 1942 everyone knew we had 48 chromosomes.
In 1943 everyone knew we had 48 chromosomes.
In 1944 everyone knew we had 48 chromosomes.
In 1945 everyone knew we had 48 chromosomes.
In 1946 everyone knew we had 48 chromosomes.
In 1947 everyone knew we had 48 chromosomes.
In 1948 everyone knew we had 48 chromosomes.
In 1949 everyone knew we had 48 chromosomes.
In 1950 everyone knew we had 48 chromosomes.
In 1951 everyone knew we had 48 chromosomes.
In 1952 everyone knew we had 48 chromosomes.
In 1953 everyone knew we had 48 chromosomes.
In 1954 everyone knew we had 48 chromosomes.

---

* Chromosomes come in pairs in humans. Biology is so much more complicated than mathematics. It seems there are always exceptions in biology (and in English, history, and sociology). How many exceptions do you know to the math rule that 2 + 2 = 4?

Humans are **diploid**. Their chromosomes come in pairs. Some plants are **triploid**. Their chromosomes come in triplets. And some plants are **tetraploid** (4). And some are **hexaploid** (6) and some are **octaploid** (8).

Chapter Forty-three   Six Words

In the early 1950s, T. C. Hsu found a neat way to scatter out the pile of chromosomes so that they were much easier to count. Did he count them? No. Everyone knew that we have 48 chromosomes.

Within about three years after Hsu invented his spread-out-the-pile-of-chromosomes trick, one guy in the United States (Albert Levan) and one guy in Sweden (Jo Hin Tijo) had nothing better to do, so they counted the number of chromosomes in human cells. They must have felt really dumb—all they could find were 46 chromosomes. Twenty-three pairs.

Each of them, Levan and Tijo, must have felt like they stood alone in the world. What do you do when the whole world says 48 chromosomes, and you count only 46? Do you get your eyes checked?

Do you make sure you can count straight? When everyone else is giving "High Fives," have you been offering them "High Fours"?

Beyond doing the obvious things like taking more cell samples and counting again, what can you do? The big temptation is just to shut up. The whole world says there are 24 pairs. Who are you to say that the world can't count?

You open your mouth and say, "Twenty-three pairs," and bad things can happen. You may get shunned by your coworkers. You may lose your job.

If history is taught poorly, all you do is learn a bunch of dates and the names of battles. If it is taught well, you experience the aloneness of individuals who dared to say "Twenty-three pairs." It has happened in every field of human endeavor—not just in genetics.

✔ In the world of art: over the centuries there have been artists who did things in an entirely different way than all the other artists around them. They were often ridiculed.

✔ In the world of mathematics: about a hundred years ago, Cantor counted the number of natural numbers {1, 2, 3, 4, . . .} that there are. He came up with an answer. Then he counted bigger infinities. He proved

Chapter Forty-three   Six Words

that there were an infinite number of infinite numbers—all different from each other. Other mathematicians told him he was nuts.

✔ In the world of religious thought: beatings, beheading, burning at the stake have awaited some individuals who said "23" when everyone else said "24."

Today . . . everyone knows that there are 23 pairs of chromosomes in every normal human body cell.*

Your Turn to Play

1. From the footnote below, "1 in 35" is what percentage? Round your answer to the nearest percent.
2. In this Chapter 43, we were going to work with six words (locus, site, DNA, chromosome, gene, and allele). We did one of them. What is one-sixth as a percent?

. . . . **COMPLETE SOLUTIONS** . . . .

1. $\frac{1}{35}$ = $35\overline{)1.000}^{0.028}$ = 0.028 = 2.8% ≐ 3%

2. There are nine conversions that you were asked to memorize in Chapter 31 of *Life of Fred: Decimals and Percents*. There were two other conversions I mentioned: $\frac{1}{6}$ = 16⅔% and $\frac{5}{6}$ = 83⅓%. At that time, we didn't add them to the list of nine conversions, because we wouldn't have nine conversions any more. If you have any extra unused brain cells, please add these two conversions today to your memory bank. For the rest of the Life of Fred series, I will assume you know the eleven conversions.

★ Babies are sometimes born with an extra 21st chromosome. They have 47 chromosomes instead of 46. They have Down syndrome. Most have significant learning disabilities. About half of kids with DS have heart defects. About half have vision and hearing problems.
    The frequency of DS varies significantly with the mother's age. For thirty-year-old mothers, 1 in 900 conceptions. For forty-five-year-old mothers, 1 in 35.

Chapter Forty-four
Five Words

Fred pulled his microscope off a shelf in his office, and popped one of his cells into the microscope. Using Hsu's spread-out-the-pile-of-chromosomes trick, he had a clear view of the 23 pairs of chromosomes.

Thankfully, the biologists have figured out an easy-to-memorize naming system for the twenty-three pairs of chromosomes.

They call the first pair. . . . . 1.
They call the second pair. . . . . . . 2.
They call the third pair. . . . . 3.

You have to admit, that's brilliant.

When Fred looked at his first pair of chromosomes, they looked virtually identical. Using just a microscope he couldn't see the individual genes on chromosome #1. If he could, he would have noticed slight differences (recall dominant and recessive genes).

The chromosome #2 pair also looked alike. A nice pair of number two chromosomes, Fred thought to himself.

When he got to the last pair (chromosome #23), he gasped. One of the pair was a regular-sized chromosome, and the other one was just a little runt. I'm a mutant!!! Fred thought. No wonder I'm only 36 inches tall. No wonder I've got such a funny nose. No wonder my eyes look like dots. No wonder my head isn't exactly round. (an understatement)

For once in his life, Fred didn't faint when he received bad news. He hopped off his chair, put the microscope back on the shelf, and raced out the door.

He flew past the vending machines in the hall, down two flights of stairs, and ran to the Theophilus Painter Memorial biology building.

All he could think of was what the newspapers would say:

Chapter Forty-four Five Words

The KITTEN Caboodle

The Official Campus Newspaper of KITTENS University

Friday noon Edition 10¢

Mutant Life Form Discovered on Campus!

KANSAS: Our University president has announced the discovery of what he calls "a severely mutated life form" on our campus. He urged the campus community not to panic. He said, "This is probably not an alien, but just an unfortunate human being with extreme damage to his chromosomes."

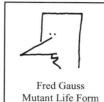

Fred Gauss
Mutant Life Form

Police Surround Math Building

In the interests of public safety, the campus police have cordoned off the math office of Professor Gauss. (continued on p. 23)

Student Interviewed

"I always thought that Gauss guy looked kind of funny."

advertisement

You can never be too safe!℠

Set of Six Rockets with Nuclear Tips
$2399.

Our motto: With mutants, you never know when you may need to go nuclear.

Fred entered the back door of the Theophilus Painter Memorial biology building. He didn't want to be seen.

"Hi, Professor Gauss," said Marie. She was one of Fred's students in his beginning algebra class. "What brings you to the biology building?"

"It's . . . could I speak with you in private?" Fred whispered.

She motioned him into her office.

After she shut the door, Fred explained, "I think I've got a broken chromosome. All my first twenty-two pairs are nicely paired up. But it's my twenty-third pair. One of them is regular size, but the other one is all wrecked up and small. Could you take a look at my number 23 under the microscope? How bad is it? Can it be fixed?"

Chapter Forty-four Five Words

Marie started to giggle. "It can't be fixed, but it's a fairly common condition. That big #23 chromosome is called an X chromosome. That little runty #23 is called a Y chromosome."

Fred didn't know what to say. Finally, he asked, "Do you have that funny X and Y chromosome combination?"

Marie burst out laughing. That didn't help Fred's mental state at all. "No one has ever accused me of having an X and a Y chromosome. I own two full-sized X chromosomes."

Fred felt a deep sense of shame when he learned that Marie has two big X chromosomes, and he was stuck with only one X and a puny Y chromosome.

Marie explained to Fred that everyone in the biology lab knew that Fred had a big X and a little Y chromosome. They didn't have to stick one of Fred's cells under a microscope to see that.

"Does it really show that much?" Fred asked.

"Yes it does." Marie went to the vending machine in the lab and brought back a chocolate milkshake for Fred. She knew that she had a lot of explaining to do.

She began, "Did your parents ever tell you about the Facts of Life?"

"I know about the facts of life," Fred answered. "Everyone knows things like two plus two equals four."

"Did your mother ever tell you where you came from?"

"Once, she mentioned that the stork brought me. That seemed kind of strange since I've never seen storks in our neighborhood. Maybe they are like Santa Claus. They are hard to spot."

"Did your mom ever tell you how parents sometimes have sons and sometimes have daughters?"

"I never thought about that," Fred said. "I guess the parents decide and then tell the stork which kind to bring."

Marie knew she had a lot of misinformation she needed to clear up. Fred's milkshake was melting in his hands as Marie explained: "If you look at chromosome #23, some people are XX and others are XY. If you are XX, then you are a girl. If you are XY, then you are a boy. If your genotype is XX, then your phenotype is female."

    Fathers have genotype XY. Mothers have genotype XX. There are two possibilities:
Father contributes X and the mother contributes X. Kid is XX. (daughter)
Father contributes Y and the mother contributes X. Kid is XY. (son)

Your Turn to Play

1. In Chapter 43, we were going to work with six words (locus, site, DNA, chromosome, gene, and allele). We only did *chromosome*. In this chapter, we were still working on *chromosome*. We still have five-sixths of the list to go. Express five-sixths as a percent.

2. Solve $3x + 7 + 2x = x + 29 + 6$

3. $(4\frac{1}{2})^3$

.......**COMPLETE SOLUTIONS**.......

1. $\frac{5}{6} = 83\frac{1}{3}\%$

2.
Original equation	$3x + 7 + 2x = x + 29 + 6$
Combine like terms on each side	$5x + 7 = x + 35$
Subtract x from both sides	$4x + 7 = 35$
Subtract 7 from both sides	$4x = 28$
Divide both sides by 4	$x = 7$

3. $(4\frac{1}{2})^3 = (\frac{9}{2})^3 = \frac{729}{8} = 8\overline{)729}\ \text{91 R1} = 91\frac{1}{8}$

Chapter Forty-five
Five Words: Locus, Site, DNA, Gene, Allele

The chocolate milkshake had completely melted. He put the cup on Marie's desk. Putting it in his pocket "for later" was not really an option. When Marie sat in Fred's beginning algebra class, he explained to her how to solve equations like $3x + 7 + 2x = x + 29 + 6$. Now she was explaining to him all about how parents pass along their genotypes to their children.

"So I have 23 pairs of chromosomes," Fred said. He was trying to encourage Marie to keep talking. "What's DNA then?"

Marie began: "Each one of your 46 chromosomes has a long skinny molecule inside it. It's a DNA molecule. It looks like two pieces of thread twisted together.

"The gene that determines your eyelash length is located in a particular chromosome and at a particular place in the DNA in that chromosome.* That location is called the **locus** of that gene. For some genetic diseases, we have been able to find the locus of the mutated gene that causes the disease.

"Cystic fibrosis is on chromosome #7. Tay-Sachs disease is on #15. Color blindness is on the X chromosome."

Marie drew on the whiteboard in her office.

Some chromosome — *The DNA* — *the locus of some gene*

"The structure of that long skinny molecule of DNA was finally figured out in 1953. A Nobel prize was awarded to the guys (Watson and Crick) who figured it out.

∗ And since chromosomes come in pairs, it would be more accurate to say that the gene for eyelash length is located at a particular place on the DNA of each of the pair of chromosomes.

Chapter Forty-five Five Words: Locus, Site, DNA, Gene, Allele

"Every* living thing has got these DNA molecules. Every sheep, every fish, every bird." Marie drew some examples on the whiteboard.

"The long skinny DNA molecule is where the information is stored that determines whether your body will be that of a boy or of a baboon. Watson and Crick figured out that the DNA molecule looks like two pieces of thread twisted together, and each thread looked like a string of beads. There are four kinds of beads: A, G, T, and C.

AGTCCCGATCGTTAAGACACCAT...

"These beads are called **nucleotides**. Your DNA consists of about three billion nucleotides. The **Human Genome Project** was started in 1990 with the goal of finding out the sequence of A's, G's, T's, and C's in our DNA. It was essentially completed in 2003."

"Wow. This is recent stuff," Fred commented. "But whose genome is it? We all have slightly different genes."

Marie explained, "All our human genomes are 99.99% alike, so it didn't matter in the Human Genome Project whose DNA they analyzed. They used a bunch of anonymous volunteers."

Your Turn to Play

1. On the previous page of this book, there are 1,172 letters. There are about three billion nucleotide letters in your genome. Roughly, how many pages of *Life of Fred* would it take to spell out your genome?

2. $4\frac{3}{4} \div 5\frac{1}{8}$

★ "Every" is a very dangerous word in biology. Every time you say *every* in biology, there seems to be some exception that shows up. But in explaining to Fred, who is a beginning student of biology, *Every living thing has DNA molecules* is at least 99.9% correct.

3. Suppose the Human Genome Project had worked on your DNA and published the list of your three billion nucleotides: AGTTCCGA.... Suppose they also did that for someone else and that person's list was identical to yours. Who would that other person be?

4. (Continuing the previous question) Suppose that other person's list differed from yours by 0.01% of the nucleotides. Who would that other person be?

·······COMPLETE SOLUTIONS·······

1. The conversion factor is $\dfrac{1 \text{ Life of Fred page}}{1172 \text{ letters}}$

$$3{,}000{,}000{,}000 \text{ letters} \times \dfrac{1 \text{ Life of Fred page}}{1172 \text{ letters}} \approx 2{,}560{,}000 \text{ pages.}$$

(≈ means "approximately equal to")

If a *Life of Fred* book contained 256 pages, then

$$2{,}560{,}000 \text{ pages} \times \dfrac{1 \text{ Life of Fred book}}{256 \text{ pages}} = 10{,}000 \text{ books.}$$

If a 256-page book were $\tfrac{3}{4}$ inches thick, then

10,000 books × 0.75 = 7500 inches tall.

$$7500 \text{ inches} \times \dfrac{1 \text{ yard}}{36 \text{ inches}} \approx 208 \text{ yards.}$$

So in 13 years, the Human Genome Project "wrote" two football field's worth of books. And this information is packed into each cell in your body.

2. $4\tfrac{3}{4} \div 5\tfrac{1}{8} = \dfrac{19}{4} \div \dfrac{41}{8} = \dfrac{19}{4} \times \dfrac{8}{41} = \dfrac{19}{\cancel{4}_1} \times \dfrac{\cancel{8}^2}{41} = \dfrac{38}{41}$

3. That person would have to be your identical twin.

4. Since all human genomes are 99.99% (or more) alike, it might be anyone on the planet (except your very close relatives).

Chapter Forty-six
Entre Nous

Just between you and me—that's what entre nous means—there are several little items that we should look at before you head on to the Final Bridge in this book. [*entre* is pronounced just like a menu *entre*. *Nous* is pronounced NEW.]

Yes, I, your reader, noticed that you didn't stick a Bridge after Chapter 42.

I'll make up for that in the Final Bridge, which will cover everything in this book.

I also noticed that in Chapter 43, you were going to define six words—chromosome, DNA, gene, locus, allele, and site—and here we are in Chapter 46, and you have defined only four of those words.

I was having too much fun. Fred just *had* to get over to the Theophilus Painter Memorial biology building in order to find out about his puny Y chromosome.

While we are in this "interrupt mode," I have one question from three pages ago. You wrote that color blindness is a mutation on a gene on the X chromosome. Everyone knows that color blindness is much more common in men than in women. Shouldn't you have written that it is a mutation on the Y chromosome? Only men have Y chromosomes.

Look at it this way. Women possess two nice big X chromosomes. If one of those two X chromosomes has a mutation in the gene that controls color recognition, their other X chromosome has a good gene in that same locus. With men, there is no corresponding backup gene on their little Y chromosome. "Knock out" that gene on their X chromosome, and they are color-blind.

✱ ← a dingbat

I've placed that dingbat here to indicate a break. I've got only about five pages left before the Final Bridge. If I may proceed without too much interruption. . . .

I won't say another word.
Thank you.
Unless I think of something.
Thank you.

Chapter Forty-six   Entre Nous

But . . .

Yes.

What's a dingbat?

It's a piece of ornamental type like ❖ or ✽ or ✪ that is used as a separator or as a border.

✻✻✻

So suppose we are looking at one of the two threads of a DNA molecule.

AGTCCCGATCGTTAAGACACCATTAGGACGCTAACACAGGTCACTATTAGAGATACCGGGATCG...

Genes are a particular stretch along this thread.   This ATTAGAGATA might be a gene. Where it is located on this DNA is its locus. The ATTAGAGATA are the nucleotides of the gene.

The position of one particular nucleotide in a gene is called a **site**. There is a T in the second site of ATTAGAGATA.

A **point mutation** occurs when a nucleotide at one particular site is changed to another nucleotide.

If ATTAGAGATA had a point mutation, it might change into ACTAGAGATA.

Some genes have two different forms. For eyelash length, there is a form for long eyelashes and a form for short eyelashes. These are not two different genes. These are two different alleles. **Alleles** are different forms of the same gene.

I did it! It took me four chapters, but I finally got the six words defined—chromosome, DNA, gene, locus, allele, and site.

You mean I know biology now? That wasn't that hard. There was lots of funny terminology, like the difference between a gene and an allele, but I think I've got it all mastered. Bring on the Final Bridge. I'm ready.

I'm not. I think I should mention the **Nine Hairy Complications**.

No. No. No. No. No.

Yes. Yes. Yes. Yes. Yes. The list is on the next page.

I'm not turning the page!

Chapter Forty-six   Entre Nous

Nine Hairy Complications of Biology

1. With eyelash length, there are two forms of the gene (called alleles). With many genes there are more than two alleles.

2. For humans, there are about 30,000 genes. Thirty thousand stretches of nucleotides that "do something." For example, we have found the gene on chromosome #7 in which a point mutation will cause cystic fibrosis. The Human Genome Project has given us a list of the 3,000,000,000 A's, G's, T's and C's, but hasn't told us where all the genes are or what their functions are.

   Somewhere between 90 and 97% of our DNA is not genes, but just seemingly random strings of nucleotides. It is called (by the biologists) **junk DNA**.* In one of the chromosomes of *Drosophila* (the fruit fly), there is the sequence AGAAG that is repeated about 100,000 times.

   This junk DNA is a hairy complication for which no one has yet come up with an explanation.

3. The whole list of A's, G's, T's and C's for the human genome is now on the Internet. A phone book is much more interesting to read.

4. The viruses that cause the flu mutate a lot faster than worms, butterflies or hippos. That is why each year they have to create a new flu shot. Last year's hippos look a lot like this year's hippos.

5. There are all kinds of mutations, not just point mutations. There is chromosomal duplication (an extra chromosome) such as in humans with Down syndrome. Some chromosomes break. Some chromosomes get flipped around (inverted). Some viruses (called retroviruses) stick themselves into your DNA, and you may end up passing your altered genome along to your kids. With **transposon insertions**, genes move from one locus in your genome to another.

∗ Mathematicians don't have any "junk numbers." It's just that we use some numbers less frequently than others. For example, it is a virtual certainty that no other human has ever written this number before: 9887239629696600328655173.

Chapter Forty-six   Entre Nous

With bacteria, things get even a little more weird. One bacteria can walk up to another* and trade genes—like kids trade baseball cards. "I'll trade you one long eyelash gene for a gene that will protect me against penicillin." With that trick, bacteria can quickly develop a resistance to particular antibiotics.

The drug companies come out with a new antibiotic that does a wonderful job of killing bacteria that cause some nasty infection such as tuberculosis. Zillions of those bad bacteria die. Then one bacterium experiences a mutation in one of its genes that allows it to survive the new antibiotic. It grows, it multiplies, and it shares its gene with other bacteria. The "new antibiotic" becomes an old antibiotic as it joins the list of drugs that don't work very well anymore.

6. Back in Chapter 21, we discussed traits that are polygenic—traits like intelligence that are affected by more than one gene. To complicate things, there are genes that are **pleiotropic**—single genes that have multiple effects. The point mutation on chromosome #7 that causes cystic fibrosis affects many parts of the body: the chemicals that the body uses in digestion are poorly secreted, thick mucus develops in the airways, etc.

Small lesson in Greek. Pleiotropic = *pleio* + *tropic*. *Tropic* means to change. *Pleio* (also spelled pleo and plio) means "more." [PLY-oh-TROW-pick where TROW rhymes with throw]

7. Your chromosomes come in nicely matched pairs (except in men where chromosome #23 has the big X and the little Y). That gives you two genes for almost everything. When the monk Gregor Mendel did his work in his garden with peas, he got lucky. When he looked at smooth seeds versus wrinkled seeds, he found that smooth seeds were the allele that was dominant over wrinkled. When he looked at yellow seeds versus green seeds, he found that the yellow allele was dominant over the green. Long stems were dominant over short stems. Inflated pea pods were dominant over constricted pods.

Mendel happened to choose traits where one allele was completely dominant over the over. It made his math so much easier.

✶ Assuming, of course, that bacteria have feet.

Chapter Forty-six   Entre Nous

Our seventh **Hairy Complication** occurs when one allele has **incomplete dominance** over another allele. In snapdragons, the red allele (R) is only partially dominant over the white allele (w).

If R were completely dominant over w, then both RR and Rw would result in red flowers. Since R is only partially dominant over w, Rw results in pink flowers. This might have driven Mendel nuts.

8. How about **codominant alleles**? Both alleles are fully expressed. One of the most common examples of codominant alleles is in blood types. The gene for blood type comes in three different alleles. (See **Hairy Complication** #1 above.) Biologists call them I^A, I^B and i. The i allele is recessive, so if you have I^A and i, your blood type is A. If you have I^B and i, your blood type is B. If you have i and i, your blood type is O.

The fun comes if you have the alleles I^A and I^B. They are codominant, and both get expressed equally. Your blood type is AB.

9. Your genes directly affect your phenotype (= what you look like). But your environment can affect your phenotype also.

Example: Your genes can affect how tall you will be, but your diet can also have an effect. If you want to effect* a change in your adult height, it's fairly tough to switch your genome, but eating a good diet as a kid, with adequate protein and calcium, may help.

Example: Your genes may dispose you to have high intelligence, but not reading and, instead, watching a lot of daytime television, can make you as dumb as dirt.

One favorite topic of conversation in some circles is whether changes you make in your phenotype will alter your genotype. If you get big muscles lifting weights, will that give you more muscular sons? The answer to that lies in a simple observation: For thousands of years, Jewish boys

---

* Affect . . . effect . . . effect. Some people actually choose to become English majors in college. Maybe they are affecting ("to pretend or feign") a desire to write long term papers and deal with the idiosyncrasies of the English language.

Feign does not obey the *i before e, except after c* rule.

245

Chapter Forty-six Entre Nous

have been circumcised. If altering one's phenotype could alter one's genotype, you might expect nowadays that some Jewish boy babies would be born pre-circumcised.

"You didn't drink your milkshake," Marie said to Fred. "Is there something else you like to drink?"

The main thing that Fred has drunk over the years was Sluice—that soft drink that has so much sugar in it that spoons float in it.

Before Fred could answer, Marie said, "I'm a biology major, and the one favorite game that biology majors love to play at parties is Name Your Favorite Mutation.

"My favorite mutation," she continued, "was first discovered by Ephraim Wales Bull. In the fall of 1843, he planted over 20,000 wild fox grape seedlings. Six years later, those seedlings fruited. BINGO! One of those plants—one out of 20,000—was a mutation that had very large, sweet fruit. Ephraim had hit the jackpot.

"He was doing his work in Concord, Massachusetts. You'll never guess what he called these new blue grapes.*

"In 1869, a guy named Thomas Welch turned some of these Concord grapes into juice, which was used in his church for communion. You'll never guess what that grape juice is called.**"

Fred continued his song, "Unbroken bones are fine with me. . ." as he headed back to his office.

---

* Concord grapes.

** Oh please. That question is too easy. If you don't know the answer, ask your kid sister the name of the *grape juice* invented by *Welch*. She'll know.

THE FINAL BRIDGE

You, dear reader, now know more biology than the average American adult.

If you don't believe me, ask the average person on the street where the mass of a two-ton tree comes from. Or ask them what is the difference between a gene and an allele.

This Final Bridge is your chance to show that you have mastered the material of *Life of Fred: Pre-Algebra 1 with Biology*. And you will also learn how Joe attempted to pay his electric bill.

This Final Bridge is a little different than the previous Bridges:
❧ There are 21 questions.
❧ Get 17 right and you pass.

On the next page is your Final Bridge. As usual, you have five tries.

The Final Bridge

first try

Goal: Get 17 or more right and you graduate.

1. A million, written in numerals, is 1,000,000. What is a trillion in numerals?

2. Being stout, having stress, drinking spirits (alcohol), and eating too much salt can increase your chances of developing hypertension. What is the common name for hypertension?

3. Joe needed some money to pay his electric bill. He held a garage sale and put up a sign that read: "Fish hooks .25¢ each." If Joe really meant what he wrote, how many fish hooks could you buy for a quarter?

4. One of Joe's fish hooks had a little bait left on it. When he put that fish hook out to sell at his garage sale, 7 grams of ants swarmed all over that hook. How many ants was that? (783 ants weigh one gram.)

5. Joe headed to the hardware store and bought a piece of lumber so that he could smash the ants. The lumber, of course, came from a tree. Much of that piece of lumber is carbon. Where did the tree get that carbon from?

6. What's an allele?

7. Joe put out two tables for his garage sale. On the first table, he had his fishing hooks and his old socks. As a set, these could be written {hooks, socks}. On the second table, he had his kindergarten class notes and a garlic plant. This second set = {notes, garlic}. Are these two sets disjoint?

8. In the first 2½ hours of Joe's garage sale, he sold nothing. He was one-third of the way through the hours he had allotted for his garage sale. What is the total number of hours he had allotted?

(continued on next page)

249

The Final Bridge

first try (continued)

9. When Betty came to Joe's garage sale, she looked at the old pair of socks that were on Joe's first table. She noticed that they didn't quite match each other. Joe's sign read: "Two socks $90." She frowned and said, "That's a lot of money for two unmatched used socks."*

Joe answered, "But my electric bill is $90, and I need the money."

"But who would pay you $90 for some old socks?" Betty asked. "The amount of your electric bill won't affect whether someone would be willing to pay you $90. I suggest you make your price a little smaller. Maybe 1% of your current price."

What new price was Betty suggesting?

10. Joe announced, "I know someone who would pay $90 for my old socks, if she knew my electric bill was $90."

"Who in their right mind would do that?" Betty exclaimed.

"My mother, but she's in Idaho."

"But this is Kansas," Betty said.

"Oh," said Joe. The thought that Joe's mom would not be coming to his garage sale had not occurred to him. Joe ran inside and got a cardboard box.

It was 4 feet wide, 6½ feet long and 2 feet deep. What is the volume of that box?

11. Joe tossed the socks into the box. They occupied 0.78% of the box. What percent of the box was empty?

(continued on next page)

---

* There is no such word as alot. There is a verb *allot*, which means to set something apart—as in the hours that Joe allotted for his garage sale.

The Final Bridge

first try (continued)

12. Joe wrote, "**Mom, Idaho**" on the outside of the box. He remembered to put a stamp on the box. Somehow, he managed to crush his package into the mailbox located at the end of his street. The postal service in Kansas receives an average of 7.8 packages each day with insufficient addresses. How many does it receive in six days?

13. Take your choice: Identify any one of these three parts of the body: (1) aorta, (2) capillaries, or (3) pinna.

14. Solve $44 + 5x = 28 + 8x - 11$

15. $5\frac{3}{4} \div 7\frac{5}{12}$ (Remember to reduce your answer if possible.)

16. The skeleton equation for burning ethylene in air is
$$C_2H_4 + O_2 \rightarrow CO_2 + H_2O$$
Balance that equation.

17. There is a gene in humans that controls whether you can curl the sides of your tongue. (There really is.) It comes in two alleles, which we'll call T and t. T = can curl. T is dominant over t. t = can't curl.

   If your father can curl his tongue, but neither you nor your mother can, what is the genotype of your father?

18. Change $\frac{1}{6}$ into a percent.

19. Solve $\frac{3}{x} = \frac{83}{7}$ (Hint: Use cross multiplying.)

20. Joe phoned his mom in Idaho and asked her if she had received the box he had sent her.

   "When did you mail it, son?" she asked.

   "About 20 minutes ago," Joe answered.

   How fast would that package have to travel in order for it to get from Joe to his mom in 20 minutes? (It's about a thousand miles between them.) Give your answer in miles per hour.

21. $\frac{1}{2} - \frac{1}{4}$

The Final Bridge

second try

Goal: Get 17 or more right and you graduate.

1. Is Fred an autotroph?
2. Find two numbers that make the equation $x^2 = 2x$ true.
3. The skeleton equation for burning ammonia, NH_3, is
$$NH_3 + O_2 \rightarrow NO + H_2O$$
Balance that equation.
4. Joe's mom looked on a map of the United States. She measured the distance from Joe's apartment near KITTENS University to her home in Idaho. It was 4½". The map said that 2½" on the map corresponds to 500 miles. How far apart are Joe and his mom? (Hint: Use a conversion factor.)
5. You need 17 (or more) right in order to pass this Final Bridge. There are 21 questions. Rounded to the nearest percent, what percent right do you need to pass?
6. [fill in two words] In any circle, π is defined as the _____ divided by the _____.
7. When they were talking on the phone, Joe's mom told him, "I'm worried that you don't eat right. I'm going to send you two dozen eggs."

    She mailed them. When they arrived at Joe's apartment, he found that seven-eighths of them were broken. How many were broken?
8. What is the volume of this cone? (Use 3 for π.)

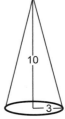

9. Looking at the gene that controls eyelash length, let L be the allele for long eyelashes and S be the allele for short eyelashes.

    L is dominant over S. Both Joe and his mom have short eyelashes. Joe's dad has long eyelashes. What is the genotype of Joe's dad?

(continued on next page)

The Final Bridge

second try (continued)

10. The fish hooks that Joe was trying to sell at his garage sale were all tangled up. Working carefully, Joe untangled 18 of them in 5 minutes. At that rate, how long would it take for him to untangle 126 fish hooks?

11. When Joe's mom had mailed the eggs, her package weighed 4 lbs, 3 oz. When it arrived at Joe's, it weighed 2 lbs., 9 oz. How much weight had the package lost?

12. Solve $30y - 24 = 28y$

13. What is the cardinal number associated with $\{4, 5, 6\}$?

14. In the first six hours that Joe held his garage sale, he had 24 people come and look at what he had for sale. Twenty-five percent of them laughed when they saw what he had for sale. How many laughed?

15. The atomic weight of hydrogen (H) is 1.
The atomic weight of carbon (C) is 12.
The atomic weight of oxygen (O) is 16.

What is the molecular weight of ethylene (C_2H_4)?

16. How many grams does a mole of water (H_2O) weigh?

17. Joe was selling his kindergarten notes at his garage sale. Most of the pages were just filled with scribbles or pictures of chickens. Joe liked to draw chickens. (It wasn't until the third grade that he learned his ABC's.)

Some sample pages from Joe's notes

Joe had 5,000,300 pages of notes. Could that be equally divided into three piles?

18. Joe originally offered his 5,000,300 pages of kindergarten notes at $1 per page. After a couple of hours, Joe noticed that no one offered him $5,000,300 for his notes. Joe wrote up a sign: **Special! 30% off**. What is the new price for the 5,000,300 pages of notes?

(continued on next page)

The Final Bridge

second try (continued)

19. Take your choice: Identify any one of these three parts of the body: (1) femur, (2) patella, or (3) retina.

20. $2^{10} = ?$

21. At the end of Joe's garage sale, he had sold 90¢ worth of fish hooks. He needed $90 to pay his electric bill. Ninety cents is what percent of $90?

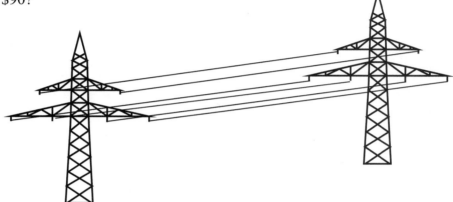

The Final Bridge

third try

1. Reduce to lowest terms $\frac{35}{42}$

   Goal: Get 17 or more right and you graduate.

2. Change $\frac{5}{6}$ into a percent.

3. Suppose there is a gene in mice that comes in two forms: R and r. Suppose that R is dominant over r. Let's say, for fun, that R is the ability to roller skate.

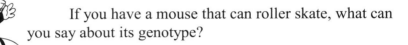

If you have a mouse that can roller skate, what can you say about its genotype?

4. A million, in numerals, is 1,000,000. What is a billion in numerals?

5. You can store most acids in glass bottles. One exception is hydrofluoric acid (HF). It eats away glass. (That's one reason why you will never see hydrofluoric acid as an ingredient in glass cleaners.) Hydrofluoric acid reacts with sodium silicate in glass:

   $$Na_2SiO_3 + HF \rightarrow H_2SiF_6 + NaF + H_2O$$

   (Na is the chemical symbol for the element sodium. They couldn't use S for sodium because S is the symbol for the element sulfur.)

   Balance that skeleton equation.

6. Change $7\frac{1}{8}$ into an improper fraction.

7. Simplify $1^{3998692369555239}$

8. Suppose a cell has a gene with the alleles U and u, where U is dominant over u. Suppose that cell undergoes mitotic division. What is (or are) the possible genotypes of the daughter cells?

(continued on next page)

# The Final Bridge

third try (continued)

9. Joe needed to get some money in a hurry to pay his overdue electric bill. He had received the Final Notice from the electric company stating that he had until 3 p.m. to pay the $90, or they would turn off his power. Joe hopped in his car and headed toward Idaho to see his mom. He drove at an average speed of 67 miles per hour for two hours and fifteen minutes. (That's 2.25 hours.) How far did he go?

10. Solve $\frac{x}{5} = \frac{31}{6}$ (Hint: Use cross multiplying.)

11. What is wrong with this story? *My father had the genotype Gg where g is a recessive lethal allele. My mother's genotype is gg. Therefore, there is a 50% chance that each of their children will have the Gg genotype.*

12. After Joe had driven for two hours and fifteen minutes, he noticed that a car was following him. It was flashing pretty red and blue lights. Its siren was wailing. And the driver was honking his horn.

    Joe pulled off the road to let this guy pass him. He wondered why this black and white car was in such a hurry.

    The police car pulled up right in back of Joe. A big man in a uniform got out of his car and came up to talk with Joe. At first, Joe thought the man in the uniform was lost and needed directions.

    After the usual pleasantries ("May I see your driver's license, registration, and proof of insurance"), he explained that Joe's car was putting out "tons of smoke."

    Your question: Is "tons of smoke" an example of (1) litotes, (2) simile, or (3) hyperbole?

(continued on next page)

The Final Bridge

third try (continued)

13. "Normally," the man in the uniform explained to Joe, "I just issue you a ticket to have you fix your car, but your smoking car is a danger to others on the road. Your car will have to be impounded. Come with me."

    The police car was almost entirely black. It was covered with soot from Joe's car. Joe drew with his finger on the hood of the police car. He drew his favorite animal.

    The policeman didn't appreciate that and told Joe to get in the car. In the backseat of the car were some orange traffic cones. They were 14" high with a 4" radius. What is the volume of one of those cones? (Use $\pi = 3$.)

14. The policeman dropped Joe off at the bus station and then drove to a car wash. Joe decided to head back to his apartment near KITTENS University rather than head on to Idaho. The bus ticket was $4. Joe only had nickels in his pocket. How many nickels did he need?

15. It would be an hour before the bus left, so Joe headed off to a fast food place and purchased a quart of Sluice. (He used his MuchoDisaster credit card since he was out of nickels.) A quart of Sluice contains 1200 calories. What is the name of the process by which the energy of the sun is converted into the "food" that Joe was about to drink? (It's one word with five syllables.)

16. A quart of Sluice has 1200 calories. How many calories are in a liter of Sluice? (1 liter = 1.057 quarts.) Round your answer to the nearest calorie. (You may use a calculator for this problem if you like.)

(continued on next page)

The Final Bridge

third try (continued)

17. There are lots of different numbers, such as 9793, 8.3, –7, $\sqrt{333}$, π, and $\frac{3969}{44444}$ but there is only one number that you are never allowed to divide by. What is that number?

18. When Joe got back to his apartment, it was 11 p.m. When he flipped on the light switch nothing happened. He felt his way to his Monster™ brand refrigerator. It was 8 feet tall, 4 feet wide and 3 feet deep. What is its volume?

19. Solve $160 = 7x + 20$

20. $3\frac{1}{8} - \frac{3}{4}$

21. $(2\frac{1}{5})^2$

The Final Bridge

fourth try

> Goal: Get 17 or more right and you graduate.

1. Is the set of all even natural numbers, {2, 4, 6, 8, 10, . . . } disjoint from the set of all natural numbers greater than 50?

2. If you combine naphthalene ($C_{10}H_8$) with oxygen, the skeleton equation is $C_{10}H_8 + O_2 \rightarrow C_8H_4O_3 + CO_2 + H_2O$
Balance that equation.

3. How many terms does 8 + 6xyz + 9 have?

4. Solve $44w - 18 = 41w$

5. Reduce to lowest terms $\dfrac{\text{one million}}{\text{one billion}}$

6. Multiple choice: Which part of the body has these two functions: Keeps bad bacteria, viruses, and toxins out of the body and protects other tissues from friction and blows?

    A) the dermis
    B) the epidermis
    C) the sternum
    D) the veins

7. When Joe opened the door of his Monster™ brand refrigerator, the light didn't go on. He said to himself, "It's probably the circuit breaker." He reached into the refrigerator and got the makings for a late evening snack: $\frac{1}{8}$ of a pound of bread, $\frac{3}{5}$ of a pound of liverwurst, $\frac{1}{4}$ of a pound of jelly. How much will his sandwich weigh?

8. Joe never has a meal without dessert. In the freezer he found some Sluice Bars. Each one was $7\frac{2}{5}$ ounces. He ate three of them. How many ounces did he eat?

(continued on next page)

259

The Final Bridge

fourth try (continued)

9. After eating his liverwurst-and-jelly sandwich and three Sluice bars in the dark, Joe felt his way to the bathroom to brush his teeth. (Years ago, when Joe had visited the dentist, he was told, "Only floss the teeth you want to keep." But there never was any floss in his medicine cabinet, so he always skipped flossing.)

   Joe figured out a way to save money. He would only use 0.07 grams of toothpaste each day. His toothpaste tube was 221 grams. How long would a toothpaste tube last? Round your answer to the nearest day.

10. Where a gene is located on a DNA molecule is called its locus. What is the particular position of a nucleotide in a gene called?

11. If a gene has two alleles, is one of them always completely dominant over the other?

12. $x^{30}x^4 = x^?$

13. Darlene had given Joe a special drinking glass. It was in the shape of a cone. Joe liked to use it to rinse out the 0.07 grams of toothpaste. The cone was 4 inches tall and had a *diameter* of 4 inches. How much water could it hold? (Use $\pi = 3$.)

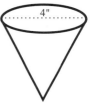

14. $(\frac{3}{4})^3$

15. Is your skin part of your integumentary system?

16. Joe wasn't very good at rinsing out his mouth. Of the 0.07 grams of toothpaste, he only managed to rinse out 60% of it. How many grams of toothpaste did he leave in his mouth?

17. If a gene has only two alleles, such as L and S, then there are three possible genotypes: LL, LS, and SS.

    If a gene has three alleles, such as A, B, and O, how many genotypes are possible?

18. If cells can't use up all of the energy they receive, some of it is put into long-term storage for use when food may become scarce. What is this storage called?

(continued on next page)

# The Final Bridge

fourth try (continued)

19. Multiple-choice: People are most likely to be at the basal metabolic rate when . . .

    A) they are doing crossword puzzles

    B) they wake up in the morning

    C) they are at maximum exertion such as running up a hill and wrestling with an alligator while solving math equations in their heads

    D) they are dead

20. Joe wandered through his dark apartment and headed to bed. Beside his bed was a big bowl of candy that weighed 17.3 pounds. Joe always liked to have a little "snack-treat" (as he called it) before heading off to sleep. After he removed 1.7 pounds of candy, how much did the bowl weigh?

21. Joe had a funny dream. He dreamed that a man from the electric company came and presented him with a bill for $90 and then told Joe that he could multiply that bill by any number he wished.

    Joe told the man, "Two," since that was Joe's favorite number.

    The bill became $180.

    The man presented Joe with a bill for $180, and Joe said, "Two" again.

    How many times would Joe say "Two" in his dream before the bill was more than a million dollars?

The Final Bridge

fifth try

Goal: Get 17 or more right and you graduate.

1. Is $2\pi > 6$?

2. Combining C_2H_4O with hydrogen gas gives rubber (C_4H_6) and water. The skeleton equation is $C_2H_4O + H_2 \rightarrow C_4H_6 + H_2O$
Balance that equation.

3. Change $12\frac{1}{4}$ into an improper fraction.

4. Solve $\frac{27}{2} = \frac{10}{w}$

5. Change $\frac{1}{8}$ into a percent.

6. How many terms does $83xyz^4$ have?

7. The dermis makes three different things that poke through the epidermis. Name any one of those three things.

8. In 1950, everyone knew we had 48 chromosomes. How many do we really have?

9. $y^6 y^6 = y^?$

10. Solve $66 + 8x + 87 = 3x + x + 333$

11. $(0.5)^2$

12. One part of our body uses about one-fourth of the oxygen that our lungs supply to our blood stream. It weighs about three pounds. What part of our body is it?

13. What is the name of the process in which the energy of sunlight is combined with carbon dioxide and water in order to form sugars, starches, oils, and cellulose?

14. Joe dreamed that his electric bill was six million dollars. Write that in numerals.

(continued on next page)

The Final Bridge

fifth try (continued)

15. He was given one last chance in his dream to multiply his six million dollar electric bill by any number he wanted. He said, "First multiply it by ten, then take that answer and multiply it by eight, and then take that answer and multiply it by six, then take that answer and multiply it by four, then take that answer and multiply it by two, and then take that answer and multiply it by zero." What would be the final answer after doing all these multiplications?

16. Joe woke up suddenly. The room was filled with light. Joe said to himself, "Somebody paid my electric bill."

That wasn't the case. It was 10 a.m., and the sun was streaming in through his windows. He got out of bed and took care of his integumentary system (washed his skin and hair and then trimmed his nails).

He called his mom in Idaho again. No answer.

He headed into the kitchen and got out a box of his favorite cereal.* The box was 9 inches tall, 6 inches wide, and $1\frac{1}{2}$ inches deep. What is its volume?

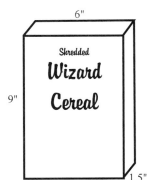

17. Joe put some cereal in a bowl. He didn't have any milk so he poured 14 ounces of Sluice over his cereal. A quart of Sluice has 1200 calories. How many calories are in 14 ounces of Sluice? (1 quart = 32 ounces)

(continued on next page)

★ Shredded Wizard cereal is featured in Chapter 21 of *Life of Fred: Calculus*

The Final Bridge

fifth try (continued)

18. While Joe ate his cereal, he read the back of the cereal box.

Joe thought that was a cool idea. His current income was the $8 that his mom sent him each week. If he could increase that $8 by 20%, what would be his new income?

19. Change 20% into a fraction. (As usual, reduce that fraction to lowest terms.)

20. Joe rubbed his eyes. "Maybe it's time for a little nap," he said to himself. As he rubbed his eyes, his fingers were closest to which of the following:

    A) his pinna, B) his retina, C) his aorta, D) his patella

21. Joe lay down* and did the thing that helped him fall asleep. He counted chickens jumping over a fence.

When Joe got to 15, he was fast asleep. Is 15 a cardinal or an ordinal number?

★ He didn't *laid down*. Joe might have laid down his spoon, but when he went to bed he lay down.

The Bridge
answers

from p. 48 — *first try*

1. There are four elements in $\{\approx, 82, \leftarrow, *\}$.
2. Yes, they are disjoint. The two sets have no elements in common.
3. $\frac{16}{17}$
4. The number 3.7 can't be a cardinal number because no set could have 3.7 members.
5. $100\% - 97\% = 3\%$. Joe is paying attention 3% of the time.
6. $9 \times 6 \times 1.1 = 54 \times 1.1 = 59.4$ cubic feet
7. $10^3 = 10 \times 10 \times 10 = 1000$
8. 25% of $16 = 0.25 \times 16 = 4$ hours per day

   Or, if you remembered that $25\% = \frac{1}{4}$

   you could have written 25% of $16 = \frac{1}{4} \times 16 = 4$
9. $v = \frac{1}{3} \pi r^2 h = \frac{1}{3} \times 3 \times 2^2 \times 12 = 48$
10. 82nd is an ordinal number.

$$\begin{array}{r} 54 \\ \times\ 1.1 \\ \hline 54 \\ 54 \\ \hline 59.4 \end{array}$$

from p. 49 — *second try*

1. $\frac{57}{8}$
2. One-sixteenth of $80 = \frac{1}{16} \times \frac{80}{1} = 5$ years old   or $16\overline{)80}$ with quotient 5
3. Since no square is a triangle, and no triangle is a square, the sets are disjoint.
4. 0
5. $\frac{88}{12}$ means $12\overline{)88}$ = 7 remainder 4, so $7\frac{4}{12} = 7\frac{1}{3} = 7.333333...$ which is not less than 7.3.

   So $\frac{88}{12} < 7.3$ is false.
6. $129 or $129. or $129.00
7. $v = \frac{1}{3} \pi r^2 h = \frac{1}{3} \times 3 \times 0.8^2 \times 2 = 0.64 \times 2 = 1.28$ cubic inches
8. 5% of $1800 = 0.05 \times 1800 = 90$ dead worms
9. $\frac{22}{103}$
10. $\frac{4}{9}$

from p. 50 — *third try*

1. Yes. The sets are disjoint.
2. $4\frac{3}{4} \times 1\frac{1}{3} = \frac{19}{4} \times \frac{4}{3} = \frac{19}{4} \times \frac{4}{3} = \frac{19}{3} = 6\frac{1}{3}$

The Bridge answers

3. $\dfrac{19}{4} \div \dfrac{4}{3} = \dfrac{19}{4} \times \dfrac{3}{4} = \dfrac{57}{16} = 3\dfrac{9}{16}$

4. 30% of 30 = 0.3 × 30 = 9 days

5. $\dfrac{5}{9}$ of an ounce

6. Since π is less than $3\dfrac{1}{7}$, it is certainly true that π < 6.

7. Since π > 3, then $π^2$ > 9. So $π^2$ < 6 is false.

8. $\dfrac{58}{7}$

9. 2×2×2×2×2×2 = 64

10. 100% − 40% = 60%

from p. 51 — *fourth try*

1. 81
2. v = ⅓ π r^2 h = ⅓ × 3 × 1.5^2 × 4 = 9 cubic feet

Or, using fractions, v = ⅓ × 3 × $(\dfrac{3}{2})^2$ × 4 = $\dfrac{9}{4}$ × 4 = 9 cubic feet

3. *First* is an ordinal number.
4. 0%
5. $\dfrac{4}{5}$ lbs.
6. $3\dfrac{1}{4} \div 2 = \dfrac{13}{4} \div \dfrac{2}{1} = \dfrac{13}{4} \times \dfrac{1}{2} = \dfrac{13}{8} = 1\dfrac{5}{8}$ lbs.
7. π is less than 4, so 2π is less than 8. Therefore, 2π > 10 can't be true.
8. $1\dfrac{1}{8} \times 6 = \dfrac{9}{8} \times \dfrac{6}{1} = \dfrac{9}{\underset{4}{8}} \times \dfrac{\overset{3}{6}}{1} = \dfrac{27}{4} = 6\dfrac{3}{4}$ magazines
9. They are not disjoint since they have elements in common.
10. $\dfrac{9}{16}$

from p. 52 — *fifth try*

1. 14 × 12 × 8 = 1344 cubic inches
2. 12% × 50 = 0.12 × 50 = 6 minutes
3. 6
4. $50 \times \dfrac{1}{5} = \dfrac{50}{5} = 10$ minutes
5. 64
6. v = ⅓ π r^2 h = ⅓ × 3 × 4^2 × 12 = 16 × 12 = 192 cubic inches
7. $\dfrac{7}{5} = 1\dfrac{2}{5}$
8. $\dfrac{1}{5}$

The Bridge answers

9. $\frac{12}{25}$

10. $\frac{4}{5} \div \frac{3}{5} = \frac{4}{5} \times \frac{5}{3} = \frac{4}{\cancel{5}} \times \frac{\cancel{5}}{3} = \frac{4}{3} = 1\frac{1}{3}$

from p. 82 — first try

1. $\frac{3}{4} \times \frac{3}{4} \times 3 = \frac{27}{16} = 1\frac{11}{16}$ cubic inches
2. 1 lb. 14 oz.

          2 lbs. 4 oz.   1 lb. 16 + 4 oz
          −   6 oz.   −   6 oz.
                               1 lb. 14 oz.

3. 300 minutes (or 5 hours)
4. 90 $V = (1/3)(3)3^2 10 = 90$
5. $\frac{3}{4}$
6. 16 lbs. $\frac{500 \text{ pages}}{1} \div \frac{5 \text{ pages}}{3 \text{ minutes}} = \frac{500 \text{ pages}}{1} \times \frac{3 \text{ minutes}}{5 \text{ pages}} = 300$ minutes
7. 128 lbs.
8. 500 × 80% = 400 pages
9. $\frac{59}{8}$ Eight times seven plus three.
10. 64   4 × 4 × 4

from p. 83 — second try

1. 368 lures   He didn't lose 92% of his lures.   0.92 × 400 = 368
2. 26¼ lbs. $\frac{15}{4} \times \frac{7}{1} = \frac{105}{4} = 26\frac{1}{4}$
3. $3\frac{1}{12}$ lbs. $3\frac{9}{12} - \frac{8}{12} = 3\frac{1}{12}$
4. True. A tenth of one cent is less than a nickel.
5. 34 of his big fishing hooks would weigh 127½ lbs., so she weighs more than 34 of his big fishing hooks.
6. 1.12 miles   r = 1.6 miles per hour   t = 0.7 hours   d = rt = (1.6)(0.7)
7. $\frac{6}{7}$ yards $\frac{36}{7} \frac{\text{yards}}{\cancel{\text{hr}}} \times \frac{1 \cancel{\text{hr}}}{6} = \frac{\overset{6}{\cancel{36}} \text{ yards}}{7} \times \frac{1}{\cancel{6}} = \frac{6}{7}$ yards
8. $\frac{37}{77}$ $\frac{33}{77} + \frac{4}{77} = \frac{37}{77}$
9. 37%   100% − 63% = 37%
10. False. 1¼ × 4 is 5, which is less than 2π, which is larger than 6.

from p. 84 — third try

1. Set A and set C are disjoint since they have no elements in common.
2. {blue, red, green, boat, balloon} (The order in which you list the elements is not important.)
3. $2\frac{7}{9}$ $\frac{1}{3} \times 8\frac{1}{3} = \frac{1}{3} \times \frac{25}{3} = \frac{25}{9} = 2\frac{7}{9}$

The Bridge
answers

4. $12.60
5. 374 cubic inches $V = (8\frac{1}{2})(11)(4) = \frac{17}{2} \times \frac{11}{1} \times \frac{4}{1} = 374$
6. $\frac{49}{64}$
7. $\frac{13}{16}$ $1 - \frac{3}{16} = \frac{16}{16} - \frac{3}{16}$
8. True. A half penny is less than ten pennies.
9. 58 worms $7\frac{1}{4} \times \frac{8}{1} = \frac{29}{4} \times \frac{8}{1} = 58$
10. 2.53 grams $\frac{66 \text{ seeds}}{1} \times \frac{2.3 \text{ grams}}{60 \text{ seeds}} = \frac{66 \text{ seeds}}{1} \times \frac{2.3 \text{ grams}}{60 \text{ seeds}} = 2.53$ grams

from p. 85 — *fourth try*

1. $3\frac{5}{12}$ lbs. $4\frac{2}{3} - 1\frac{1}{4} = 4\frac{8}{12} - 1\frac{3}{12} = 3\frac{5}{12}$
2. $5\frac{11}{30}$ feet $1\frac{1}{5} + 4\frac{1}{6} = 1\frac{6}{30} + 4\frac{5}{30}$
3. $16.60 $0.83 \times 20 = 16.6$
4. $5\frac{5}{6}$ inches $\frac{1}{3} \times 17\frac{1}{2} = \frac{1}{3} \times \frac{35}{2} = \frac{35}{6}$
5. $11\frac{2}{3}$ inches $\frac{2}{3} \times 17\frac{1}{2} =$ etc.   or $17\frac{1}{2} - 5\frac{5}{6} =$ etc.
6. $\frac{2}{9}$ $\frac{2}{9}$ is $\frac{4}{18}$ $\frac{1}{6}$ is $\frac{3}{18}$
7. $165\frac{1}{3}$ cubic miles $V = (\frac{1}{3})\pi r^2 h = \frac{1}{3}(3.1)(4^2)(10) = \frac{1}{3}(3.1)(16)(10) = \frac{496}{3}$
8. 640.08 cm $\frac{7 \text{ yards}}{1} \times \frac{91.44 \text{ cm}}{1 \text{ yard}} = 640.08$ cm
9. $\frac{56}{9}$ Nine times six plus 2
10. $(3\frac{2}{3})^2 = (\frac{11}{3})^2 = \frac{121}{9} = 13\frac{4}{9}$

from p. 86 — *fifth try*

1. $\frac{8}{125}$
2. 27% $100\% - 73\%$
3. 125 lbs. $\frac{1}{8} \times \frac{1000}{1} = \frac{1000}{8}$ $8\overline{)1000} = 125$
4. 1 lb. 12 oz.   5 lb. 3 oz. − 3 lbs. 7 oz. = 4 lbs. (16+3) oz − 3 lbs. 7 oz.
5. 4900 calories
6. The bridal magazines $\frac{1}{2}$ is $\frac{8}{16}$ which is less than $\frac{9}{16}$
7. $\frac{1}{8}$
8. $5.40/lb.   60% of $9/lb. = $5.40/lb.
9. $1\frac{1}{6}$ $\frac{7}{8} \div \frac{3}{4} = \frac{7}{8} \times \frac{4}{3} = \frac{7}{8} \times \frac{4}{3} = \frac{7}{6} = 1\frac{1}{6}$
10. zero

The Bridge
answers

from p. 114 — *first try*

1. 4%   20 = ?% of 500   Divide the number closest to the "of" into the other number.
$$\frac{20}{500} = 0.04 = 4\%$$

2. 150 pages   30% of 500 = 0.3 × 500 = 150 pages
3. 1 lb. 12 oz.   2 lbs. 7 oz. − 11 oz. = 1 lb. 23 oz. − 11 oz. = 1 lb. 12 oz.
4. x = 6
5. The two equally likely cases for the children are: L from dad, S from mom
                                                         S from dad, S from mom.
In the first case, a child with LS would have long eyelashes since L is dominant.
In the second case, a child with SS would have short eyelashes.
So 50% of the children would be expected to have long eyelashes.
6. 59¢ × 7 = 413¢ or $4.13
7. $\frac{57}{7}$ Seven times eight plus one
8. The cardinal number associated with {2, 7, 8} is the number of elements in that set. (This was defined in question one on page 48.) So the cardinal number associated with {2, 7, 8} is 3.
9. $1\frac{1}{3}$ cubic inches $\frac{2}{3} \times \frac{2}{\cancel{3}} \times \frac{\cancel{3}}{1} = \frac{4}{3} = 1\frac{1}{3}$
10. Since π ≈ 3.14, 3π would be greater than 9.

from p. 115 — *second try*

1. V = ⅓πr²h = ⅓(3)3²(7) = 63 cubic feet
2. Using a conversion factor: $\frac{70 \text{ cubic feet}}{1} \times \frac{7.5 \text{ gallons}}{1 \text{ cubic foot}} = 525$ gallons
3.   10 minutes   9 minutes 60 sec
   − 3 minutes 14 sec   − 3 minutes 14 sec
                              6 minutes 46 sec
4. $\frac{1}{12}$ → $12\overline{)1.000}^{\,0.083}$ → 8.3% which rounds to 8%
5. Retina are the photoreceptors on the back wall of the eye.
6. 20%
7. Yes.
8. $(4¼)^2 = (\frac{17}{4})^2 = \frac{289}{16} = 18\frac{1}{16}$
9. $\frac{5 \text{ ounces}}{1} \times \frac{\$0.42}{1 \text{ ounce}} = \$2.10$
10. 343

The Bridge
answers

from p. 116 — *third try*

1. 2 oz. is ?% of 1 cup   2 oz. is ?% of 8 oz.   8)2̄   25%
2. 4 inches   7 feet   6 feet 12 inches
            − 6 feet 8 inches   − 6 feet   8 inches
                                                    4 inches

3. d = rt $\frac{1.8 \text{ ft}}{1 \text{ sec}} \times \frac{7 \text{ sec}}{1} = 12.6$ feet

4. 36% of $120   0.36 × 120 = $43.20

5. $\frac{5}{6} - \frac{4}{5} = \frac{25}{30} - \frac{24}{30} = \frac{1}{30}$

6. Plants are autotrophs. They take water and carbon dioxide and add sunlight to get sugars, starches and oils. Joe is a heterotroph. He can't do photosynthesis.

7. 3 is ?% of 150   150)3.00̄ = 0.02 = 2%

8. 10%

9. There are four possible answers: { }, {5}, {8}, or {5, 8}

10. $\frac{5}{6} \times \frac{4}{5} = \frac{\cancel{5}}{6} \times \frac{4}{\cancel{5}} = \frac{4}{6} = \frac{2}{3}$

from p. 117 — *fourth try*

1. 3897 is a cardinal number, not an ordinal number.

2. $\frac{7.1 \text{ seconds}}{1 \text{ oz.}} \times \frac{16 \text{ ozs.}}{1 \text{ lb.}} = \frac{113.6 \text{ seconds}}{1 \text{ lb.}}$

3. 9 lbs. 11 oz.   160 lbs.   159 lbs. 16 oz.
                      − 150 lbs 5 oz.   − 150 lbs.   5 oz.
                                              9 lbs. 11 oz.

4. Three-fourths of a pound is $\frac{3}{\cancel{4}_1} \times \frac{\cancel{16}^4}{1} = 12$ ounces. This is heavier than the eleven-ounce ax.

5. d = rt $\frac{8 \text{ ft}}{1 \cancel{\text{sec}}} \times \frac{78 \cancel{\text{ sec}}}{1} = 624$ feet

6. 55/91

7. Pinnas are used to hold up glasses. Pinnas = ears. Another meaning of *pinna* is a fin or flipper. So in some sense, ears are flippers for the head.

8. $(3⅓)^2 = (\frac{10}{3})^2 = \frac{100}{9} = 11\frac{1}{9}$

9. 7% of $3.60 → 0.07 × 3.60 = $0.252 which rounds to $0.25 or 25¢.

10. 1/32

from p. 118 — *fifth try*

1. V = πr²h = (3.1)(1)²(2) = 6.2 cubic feet

The Bridge
answers

2. $\dfrac{6 \text{ minutes}}{1} \times \dfrac{1 \text{ hour}}{60 \text{ minutes}} = \dfrac{6 \text{ hours}}{60}$ = one-tenth of an hour or 0.1 hours.

3. d = rt   d = 50 mph × 0.1 hours = 5 miles

4. The bucket's volume is 6.2 cubic feet (from question one). Three-fourths of 6.2 is

$\dfrac{3}{4} \times \dfrac{6.2}{1} = \dfrac{18.6}{4} = 4.65$ cubic feet or $4\dfrac{13}{20}$ cubic feet

5. 2^3, which is 8, is larger than 1 or 0.

6. One-eighth was one of the nine conversions to percents that you memorized in Chapter 31 of *Life of Fred: Decimals and Percents*. $\dfrac{1}{8} = 12\frac{1}{2}\%$

7. If you get an 18% discount, you pay 82% of the price. 82% of $200 = $164.

8. $\dfrac{2}{7} + \dfrac{1}{8} = \dfrac{16}{56} + \dfrac{7}{56} = \dfrac{23}{56}$

9. $\dfrac{2}{7} - \dfrac{1}{8} = \dfrac{16}{56} - \dfrac{7}{56} = \dfrac{9}{56}$

10. $(2\dfrac{4}{5})^2 = (\dfrac{14}{5})^2 = \dfrac{196}{25} = 7\dfrac{21}{25}$ or 7.84

from p. 154 — *first try*

1. 7/8 = 87½% (This was one of the nine conversions you were asked to memorize in Chapter 31 of *Life of Fred: Decimals and Percents*.) Here is the complete list:

$\dfrac{1}{2} = 50\%$ $\dfrac{1}{8} = 12\frac{1}{2}\%$

$\dfrac{1}{3} = 33\frac{1}{3}\%$ $\dfrac{3}{8} = 37\frac{1}{2}\%$

$\dfrac{2}{3} = 66\frac{2}{3}\%$ $\dfrac{5}{8} = 62\frac{1}{2}\%$

$\dfrac{1}{4} = 25\%$ $\dfrac{7}{8} = 87\frac{1}{2}\%$

$\dfrac{3}{4} = 75\%$

2. $88\% = \dfrac{88}{100} = \dfrac{22}{25}$

3. $\dfrac{6 \text{ lbs.}}{1} \times \dfrac{16 \text{ oz.}}{1 \text{ lb.}} = 96$ ounces

4. 100% − 98% − 1% − 0.5% = 0.5%

   0.5% × 96 ounces = 0.48 ounces.

   That's less than a half ounce. The dog will starve to death.

The Bridge answers

5. $V = \pi r^2 h = (3)(25)(8) = 600$ cubic inches

6. $6x + 8 = 476$
   $6x = 468$
   $x = 78$ Each dog cost \$78.

7. $2 + 4 + 6$ has three terms. (That was covered in the footnote on page 132.)

8. $\frac{51}{5}$

9. The set $\{1, 3, 5, 7\}$ has 4 elements in it.

10. $7 \times 7 \times 7 \times 5 \times 5 \times 5 \times 0 \times 0 \times 0 \times 0 \times 0 \times 0 \times 0 \times 0 = 0$, since multiplying anything by zero gives you an answer of zero.

from p. 155 — second try

1. No. *Fifth* is an ordinal number. The cardinal numbers are the numbers that can be used to count the elements in a set. The cardinal number associated with $\{✂, ✐\}$ is 2.

2. We subtracted 3x from both sides of the equation.

3. $6 + 15w + 9w^2$. Like terms have exactly the same letters with exactly the same exponents.

4. π is approximately 3.14, so $\pi - 3$ is approximately 0.14, which is greater than 0.1. So $\pi - 3 > 0.1$ is true.

5. $\frac{600 \text{ cubic inches}}{1} \times \frac{2 \text{ ounces}}{1 \text{ cubic inch}} \times \frac{1 \text{ pound}}{16 \text{ ounces}} = 75$ pounds

6. No. We know the volume of the pennies. We know the weight of the pennies. But without a conversion factor (such as 93 pennies occupy 1 cubic inch or 8 pennies weigh one ounce) we can't determine how many pennies he has. (I have invented these conversion factors. They are not the actual values.)

7. Joe's possessions total \$960. One-eighth of 960 is \$120.

8. Dividing top and bottom by six, we get $\frac{4}{5}$

9. $\frac{4}{5} = \frac{8}{10} = 80\%$ or you could have divided $5\overline{)4.00} = 0.80 = 80\%$

10. $5\frac{2}{3} \div 2\frac{2}{3} = \frac{17}{3} \div \frac{8}{3} = \frac{17}{3} \times \frac{3}{8} = \frac{17}{8} = 2\frac{1}{8}$

from p. 156 — third try

1. $8x + 23 = 12023$
   $8x = 12000$
   $x = 1500$ Joe has 1500 pennies.

2. 1492¢ = \$14.92

3. Using a conversion factor: $\frac{1492 \text{ pennies}}{1} \times \frac{15 \text{ seconds}}{4 \text{ pennies}} = 5595$ seconds

The Bridge
answers

4. To convert 0.25 minutes into seconds, we use the a conversion factor:

$$\frac{0.25 \text{ minutes}}{1} \times \frac{60 \text{ seconds}}{1 \text{ minute}} = 15 \text{ seconds.}$$

So 93.25 minutes equals 93 minutes and 15 seconds.

5. We want 80% off of a regular price of $120. So we are going to pay 20% of $120.
20% of $120 = 0.2 × 120 = $24.

6. Those two sets have no elements in common. Therefore, they are disjoint.

7. 6% of $36 = 0.06 × 36 = 2.16   The sales tax is $2.16.

8. $2^3 = 2\times2\times2 = 8$ $3^2 = 3\times3 = 9$ It is not true that $2^3 > 3^2$.

9. Using a conversion factor: $\frac{17 \text{ square inches}}{1} \times \frac{8 \text{ fleas}}{1 \text{ sq inch}} = 136$ fleas

10. $4\frac{4}{5} \div 3\frac{4}{7} = \frac{24}{5} \div \frac{25}{7} = \frac{24}{5} \times \frac{7}{25} = \frac{168}{125} = 1\frac{43}{125}$

from p. 157 — fourth try

1. $V = \pi r^2 h = (3.1)(4)(7) = 86.8$ cubic inches

2. Using a conversion factor: $\frac{7/8 \text{ bowl}}{1} \times \frac{48 \text{ seconds}}{3/5 \text{ bowl}} = \frac{7}{8} \times \frac{48}{1} \div \frac{3}{5} =$

$\frac{7}{\cancel{8}} \times \frac{\cancel{48}^6}{1} \times \frac{5}{3} = \frac{7}{1} \times \frac{\cancel{6}^2}{1} \times \frac{5}{\cancel{3}} = 70$ seconds

3. d = rt   distance equals rate times time. $d = \frac{6 \text{ feet}}{\text{sec}} \times \frac{18 \text{ sec}}{1} = 108$ feet

4. $17.35 + $0.72 = $18.07

5. $\frac{1}{3} + \frac{3}{5} = \frac{5}{15} + \frac{9}{15} = \frac{14}{15}$ lbs.

6. 8 = ?% of 32. Divide the number closest to the *of* into the other number.

$$\frac{8}{32} = \frac{1}{4} = 25\%$$

7. 9% of 99 = 0.09 × 99 = 8.91

8. $\frac{1}{12}$ The 𝔊𝔢𝔫𝔢𝔯𝔞𝔩 ℜ𝔲𝔩𝔢 is that if you don't know whether to add, subtract, multiply, or divide, first restate the problem with really simple numbers. Instead of "How much larger is 3/4 than 2/3?" we could ask, "How much larger is 10 than 8?" Immediately, we know the answer is 2. How did we get 2? We subtracted. So "How much larger is 3/4 than 2/3?" becomes $\frac{3}{4} - \frac{2}{3}$ which then becomes $\frac{9}{12} - \frac{8}{12} = \frac{1}{12}$

9. 32 + 3x = 5x   32 = 2x   16 = x

10. 932.
   + 0.932
   ─────────
     932.932

 **The Bridge**
answers

from p. 158 — *fifth try*

1. $3x + 25 = 226$ $3x = 201$ $x = 67$ lbs.
2. The systolic reading is the upper reading. In this case, it is 16.
3. 16th is an ordinal number.
4. Two sets are disjoint if they have no elements in common. There is no animal that is both a dog and a rat, so the sets are disjoint.
5. All things on earth are either living or non-living. The union of the set of all living things on earth with the set of all non-living things on earth is the set of all things on the earth.
6. 0.25
7. $12w - 31 = 4w + 9$ $12w = 4w + 40$ $8w = 40$ $w = 5$
8. Any number divided by itself equals one.
9. $\frac{100}{3}$
10. $6 = ?\%$ of 60. You divide the number closest to the *of* into the other number.
$\frac{6}{60} = \frac{1}{10} = 0.1 = 10\%$ or you could have $60 \overline{)6.00}^{\,0.10} = 10\%$

from p. 189 — *first try*

1. $4Fe + 3O_2 \rightarrow 2Fe_2O_3$
2. 3 is what percent of 20? $20\overline{)3.00}^{\,0.15}$ 15%
3. The sum of the digits of 4,000,700 is 11. The sum of the digits of 4,000,701 is 12. Since 12 is divisible by 3, so is 4,000,701.
4. $\frac{5}{2} = \frac{x}{18}$ $(5)(18) = 2x$ $90 = 2x$ $45 = x$ You need 45 fighting scenes.
5. 39 lbs. 14 oz.   40 lbs.   39 lbs. 16 oz.
                    $-$ 2 oz. $-$ 2 oz.
                                                39 lbs. 14 oz.
6. Using a conversion factor, $\frac{12 \text{ minutes}}{1} \times \frac{1 \text{ hour}}{60 \text{ minutes}} = \frac{12}{60}$ hours $= \frac{1}{5}$ hour
7. $10 \div 0.01$ equals 1000. π is less than 4. So π^4 is less than 4^4, which is 256. So the largest of the three numbers is $10 \div 0.01$.
8. If the discount is 20%, then you have to pay 80% of the price. 80% of $7 is $5.60.
9. 3/20 $3/4 - 3/5 = 15/20 - 12/20 = 3/20$
10. 1¼ $3/4 \div 3/5 = \frac{\cancel{3}}{4} \times \frac{5}{\cancel{3}} = \frac{5}{4}$

from p. 191 — *second try*

1. 2 is ?% of 50 $\frac{2}{50}$ which is $\frac{4}{100}$ or 4%. You could also have done this by $50\overline{)2.00}$

275

The Bridge
answers

2. $\frac{3}{2} = \frac{39}{x}$ $3x = 78$ $x = 26$ words

3. The conversion factor is $\frac{1.6 \text{ pages}}{1 \text{ minute}}$ so $\frac{180 \text{ minutes}}{1} \times \frac{1.6 \text{ pages}}{1 \text{ minute}} = 288$ pages

4. $\frac{20 \text{ ft}}{1} \times \frac{1 \text{ second}}{2.5 \text{ ft}} = \frac{20}{2.5}$ seconds = 8 seconds

5. 29.7 square feet

6. $\frac{54 \text{ grams}}{1} \times \frac{1 \text{ mole}}{27 \text{ grams}} = 2$ moles

7. One mole of anything has Avogadro's number of particles. 6.0221367×10^{23} atoms

8. $2H_2O_2 \rightarrow 2H_2O + O_2$

9. $(\frac{10}{3})^3 = \frac{1000}{27} = 37\frac{1}{27}$

10. $5x + 13 + 7x = 16x + 1$ $12x + 13 = 16x + 1$ $13 = 4x + 1$ $12 = 4x$ $3 = x$

from p. 193 — third try

1. $V = \pi r^2 h = (3)(16)(18) = 864$ cubic inches

2. 67.3% $100\% - 32.7\% = 67.3\%$

3. Molecular weight of H_2O_2 is $1 + 1 + 16 + 16 = 34$, so a mole would weigh 34 grams. Four moles would weigh 136 grams.

4. Is 25 divisible by 7? No. If the sum of the digits is equal to 7, that doesn't make the number divisible by 7.

5. Using a conversion factor: $\frac{27 \text{ ft}}{1} \times \frac{12 \text{ in}}{1 \text{ ft}} = 324$ inches

6. 324 inches ÷ 10 = 32.4 inches which rounds to 32 inches.

7. $19 \times 12 + 28 \times 1 + 2 \times 16 = 288$

8. 62.3 grams

9. 37.5% or 37½% $8\overline{)3.000}$ = 0.375

10. Joe had 1½ pounds more ice cream. $3\frac{1}{4} - 1\frac{3}{4} = 2\frac{5}{4} - 1\frac{3}{4} = 1\frac{2}{4} = 1\frac{1}{2}$

from p. 195 — fourth try

1. Using a conversion factor: $\frac{5 \text{ lbs.}}{1} \times \frac{16 \text{ oz.}}{1 \text{ lb.}} = 80$ ounces

2. 0.923%

3. $V = \frac{1}{3}\pi r^2 h = (\frac{1}{3})(3)(1.5)^2(4) = 9$ cubic inches

4. $2C_{57}H_{110}O_6 + 163O_2 \rightarrow 114CO_2 + 110H_2O$ (The moral of this story is that one molecule of animal fat takes a lot of oxygen and produces a lot of carbon dioxide and water—and in the process, produces a lot of calories.)

5. $\frac{91 \text{ seconds}}{1} \times \frac{6 \text{ snores}}{13 \text{ seconds}} = 42$ snores

The Bridge
answers

6. The following whole numbers make x < 7 true: 0, 1, 2, 3, 4, 5, 6. There are seven numbers in that list.

7. $1{,}000{,}000 \div 30 = 33{,}333\frac{1}{3}$ which rounds to 33,333 calories/day.

8. x = 11

9. 39.78

```
   39.7
+  0.08
───────
  39.78
```

10. $4\frac{21}{25}$ $\left(\frac{11}{5}\right)^2 = \frac{121}{25} = 4\frac{21}{25}$

from p. 196 — *fifth try*

1. 2160 calories

2. 6' 8" = 72" + 8" = 80"   80 × 36 = 2880 square inches

3. $\dfrac{300 \text{ sq ft}}{1 \text{ gallon}} \times \dfrac{1 \text{ gallon}}{4 \text{ quarts}} \times \dfrac{1 \text{ quart}}{2 \text{ pints}} = 37.5$ sq ft/pint or $37\frac{1}{2}$ sq ft/pint

4. 4 is ?% of 20 $\frac{4}{20}$ which is $\frac{1}{5}$ which is 20%

5. $3NaHCO_3 + H_3C_6H_5O_7 \rightarrow 3CO_2 + 3H_2O + Na_3C_6H_5O_7$

6. Anything times zero always equals zero.

7. 2000 × $1.20 = $2400

8. He had 1 lb. 5 oz. left.

```
  2 lbs              1 lb   16 oz
-       11 oz      -        11 oz
─────────────      ─────────────
                     1 lb    5 oz
```

9. 4 oz is ?% of 32 oz. $\frac{4}{32}$ which is $\frac{1}{8}$ which is $12\frac{1}{2}$%

10. $\dfrac{16}{256} = \dfrac{1}{16}$

from p. 249 — *first try*

1. One trillion is 1,000,000,000,000.

2. The common name for hypertension is high blood pressure.

3. .25¢ is one-fourth of a penny. That means you could buy 4 fish hooks for a penny. For twenty-five cents, you could buy a hundred fish hooks.

4. 7 grams × 783 ants/gram = 5481 ants

5. Most of the non-water weight of a plant comes from the carbon dioxide in the air. (Chapter 15–16)

6. If a gene comes in several different forms, these forms are called alleles.

7. Two sets are disjoint if they have no elements in common. Yes, they are disjoint.

8. $3 \times 2\frac{1}{2} = 7\frac{1}{2}$ hours (Chapter 3)

9. One percent of $90 is $0.90 (or 90¢).

10. $4 \times 6\frac{1}{2} \times 2 = 52$ cubic feet.

11. 100% – 0.78% = 99.22%

12. 6 days × 7.8 packages/day = 46.8 packages

The Bridge
answers

13. Aorta is the largest artery in the body. It comes directly out of the heart. (Chapter 29)

Capillaries are the tiny blood vessels that connect the arteries to the veins. (Chapter 23)

Pinna is the visible part of the ear. (Chapter 19)

14. Original equation $44 + 5x = 28 + 8x - 11$

Combine like terms $44 + 5x = 17 + 8x$

Subtract 17 from both sides $27 + 5x = 8x$

Subtract 5x from both sides $27 = 3x$

Divide both sides by 3 $9 = x$

15. $5\frac{3}{4} \div 7\frac{5}{12} = \frac{23}{4} \div \frac{89}{12} = \frac{23}{4} \times \frac{12}{89} = \frac{23}{\cancel{4}_1} \times \frac{\cancel{12}^3}{89} = \frac{69}{89}$

16. $C_2H_4 + 3O_2 \rightarrow 2CO_2 + 2H_2O$

17. Your father must be Tt. The reasoning: Both you and your mother must be tt. Since your father can curl his tongue, he must be either TT or Tt. He couldn't be TT, because then all his children would be able to curl their tongues.

18. $\frac{1}{6} = 16\frac{2}{3}\%$

19. Original equation $\frac{3}{x} = \frac{83}{7}$

Cross multiply $21 = 83x$

Divide both sides by 83 $\frac{21}{83} = x$

20. Using conversion factors: $\frac{1000 \text{ miles}}{20 \text{ minutes}} \times \frac{60 \text{ minutes}}{1 \text{ hour}} = \frac{3000 \text{ miles}}{\text{hour}}$

21. $\frac{1}{2} - \frac{1}{4} = \frac{2}{4} - \frac{1}{4} = \frac{1}{4}$

from p. 252 — *second try*

1. Autotrophs are plants that take water and carbon dioxide, add some energy from the sun and produce sugars, starches, and oils. Fred is a heterotroph. He eats plants and turns their sugars, starches and oils into carbon dioxide, water, and energy.

2. Zero and two will each make $x^2 = 2x$ true.

3. $4NH_3 + 5O_2 \rightarrow 4NO + 6H_2O$

4. $\frac{4\frac{1}{2} \text{ inches}}{1} \times \frac{500 \text{ miles}}{2\frac{1}{2} \text{ inches}} = 4\frac{1}{2} \div 2\frac{1}{2} \times 500 = \frac{9}{2} \div \frac{5}{2} \times \frac{500}{1} =$

$\frac{9}{2} \times \frac{2}{5} \times \frac{500}{1} = \frac{9}{\cancel{2}_1} \times \frac{\cancel{2}}{\cancel{5}_1} \times \frac{\cancel{500}^{100}}{1} = 900 \text{ miles}$

278

The Bridge
answers

5. 17 is what percent of 21? Dividing the number closest to the "of" into the other number, we have $21\overline{)17.000}^{\,0.809} = 80.9\% \doteq 81\%$

6. In any circle, π is defined as the circumference divided by the diameter. (Chapter 6)

7. $\dfrac{7}{\cancel{8}_1} \times \dfrac{\cancel{24}^{\,3}}{1} = 21$

8. 90 $V = \tfrac{1}{3}\pi r^2 h = (\tfrac{1}{3})(3)(3^2)(10) = 90$

9. LS   Both Joe and his mom must have SS as their genotype. Joe's dad couldn't have SS because he has long eyelashes. Joe's dad couldn't have LL because Joe has short eyelashes. That only leaves LS as a possibility for Joe's dad.

10. Knowing that Joe can untangle 18 fish hooks in 5 minutes gives us the conversion factor. $\dfrac{\cancel{126}^{\,7} \text{ fish hooks}}{1} \times \dfrac{5 \text{ minutes}}{\cancel{18} \text{ fish hooks}} = 35 \text{ minutes}$

11. The package lost 1 lb. 10 oz.

```
   4 lbs.  3 oz.        3 lbs.  16+3 oz.
 − 2 lbs.  9 oz.      − 2 lbs.     9 oz.
                        1 lb.     10 oz.
```

12. Original equation $30y − 24 = 28y$
    Add 24 to both sides $30y = 28y + 24$
    Subtract 28y from both sides $2y = 24$
    Divide both sides by 2 $y = 12$

13. 3   The cardinal number associated with a set is the number of elements in that set.

14. Twenty-five percent of 24 is the same as one-quarter of 24, which is 6.

15. 28 C_2H_4 $2(12) + 4(1) = 28$

16. 18 grams   The molecular weight of H_2O is $2 + 16 = 18$, so a mole of it weighs 18 grams.

17. No. The sum of the digits of 5,000,300 is 8. Since 8 is not evenly divisible by 3, then neither is 5,000,300.

18. If Joe is offering a 30% discount, that means that a purchaser will have to pay 70% of the original price. 70% of $5,000,300 = 0.7 × $5,000,300 = $3,500,210.

19. The femur is another name for the thighbone. The patella is another name for kneecap. The retina are the photoreceptors on the back wall of the eyes.

20. $2^{10} = 1024$

21. 90¢ is what percent of $90?
    $0.90 is what percent of $90?
    You divide the number closest to the *of* into the other number. $90\overline{)0.90}^{\,0.01}$ $0.01 = 1\%$.

 **The Bridge**
answers

from p. 255 — *third try*

1. $\frac{35}{42} = \frac{5}{6}$ after dividing top and bottom by 7.

2. $\frac{5}{6} = 83\frac{1}{3}\%$

3. That roller skating mouse could not have genotype rr. It must either be RR or Rr.

4. A billion is 1,000,000,000.

5. $Na_2SiO_3 + 8HF \rightarrow H_2SiF_6 + 2NaF + 3H_2O$

6. $\frac{57}{8}$ Eight times seven plus one

7. 1 times itself any number of times equals 1.

8. Mitosis is the normal cell division that occurs in most cells in the body. Back in Chapter 42, we noted, "This process—mitosis—preserves the whole set of genes. If, for example, your cells contained one dominant eyelash gene, L, and one recessive eyelash gene, S, then after mitosis, each daughter cell would have LS."

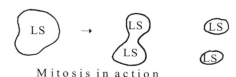

Mitosis in action

Therefore, the only possible genotype will be Uu for each daughter cell.

9. distance = rate times time. d = rt $d = \frac{67 \text{ miles}}{\text{hour}} \times \frac{2.25 \text{ hours}}{1} = 150.75$ miles

(or $150\frac{3}{4}$ miles)

10. Original equation $\frac{x}{5} = \frac{31}{6}$

Cross multiply $6x = 155$

Divide both sides by 6 $x = \frac{155}{6} = 25\frac{5}{6}$

11. It's the second sentence in this story that doesn't make sense. *My mother's genotype is gg.* No one's mother could have a gg genotype where g is a recessive lethal allele. She would have died either before birth or shortly thereafter. She could never have grown up to be anyone's mother. (See the answer to question 2a in the *Your Turn to Play* at the end of Chapter 41.)

12. "Tons of smoke" is an exaggeration. It is hyperbole.

13. 224 cubic inches $V = \frac{1}{3}\pi r^2 h = (\frac{1}{3})(3)(4^2)(14) = 224$ cubic inches

14. This could be solved in several different ways. We'll use a conversion factor.

$\frac{\$4}{1} \times \frac{1 \text{ nickel}}{\$0.05} = 0.05\overline{)4.00}^{80}$ 80 nickels

280

The Bridge
answers

15. Photosynthesis. (Chapters 15 and 16)

16. $\dfrac{1200 \text{ calories}}{1 \text{ quart}} \times \dfrac{1.057 \text{ quarts}}{1 \text{ liter}} = \dfrac{1268.4 \text{ calories}}{1 \text{ liter}} \doteq 1268$ calories/liter.

($\doteq$ means "equals after rounding off")

17. Division by zero is not permitted. (Chapter 36)

18. 96 cubic feet $V = (8)(4)(3)$

19. Original equation $160 = 7x + 20$
    Subtract 20 from both sides $140 = 7x$
    Divide both sides by 7 $20 = x$

20. $3\dfrac{1}{8}$ $3\dfrac{1}{8}$ $2\dfrac{9}{8}$
    $-\dfrac{3}{4}$ $-\dfrac{6}{8}$ $-\dfrac{6}{8}$
                                        $\overline{2\dfrac{3}{8}}$

21. $(2\dfrac{1}{5})^2 = 2\dfrac{1}{5} \times 2\dfrac{1}{5} = \dfrac{11}{5} \times \dfrac{11}{5} = \dfrac{121}{25} = 4\dfrac{21}{25}$

from p. 259 — *fourth try*

1. They are not disjoint. For example, the number 52 is in both sets.

2. $2C_{10}H_8 + 9O_2 \rightarrow 2C_8H_4O_3 + 4CO_2 + 4H_2O$

3. It has three terms. If you combined like terms, it would have fewer terms, but as it stands, it has three terms. (see the footnote on the fourth page of Chapter 25)

4. Original equation $44w - 18 = 41w$
   Add 18 to both sides $44w = 41w + 18$
   Subtract 41w from both sides $3w = 18$
   Divide both sides by 3 $w = 6$

5. $\dfrac{1,000,000}{1,000,000,000} = \dfrac{1}{1,000}$

6. B) The epidermis keeps bad bacteria, viruses and toxins out of the body and protects other tissues from friction and blows.

7. $\dfrac{1}{8} + \dfrac{3}{5} + \dfrac{1}{4} = \dfrac{5}{40} + \dfrac{24}{40} + \dfrac{10}{40} = \dfrac{39}{40}$ pounds

8. $7\dfrac{2}{5} \times 3 = \dfrac{37}{5} \times \dfrac{3}{1} = \dfrac{111}{5} = 22\dfrac{1}{5}$ ounces.

9. We will use a conversion factor: $221 \text{ grams} \times \dfrac{1 \text{ day}}{0.07 \text{ grams}} = 0.07\overline{)221.000}$
   $7\overline{)22100.0} = 3157.1 \doteq 3157$ days. $\dfrac{3157.1}{7)22100.0}$

10. The position of a particular nucleotide in a gene is called a site.

11. No. (The two alleles could be codominant, or one allele could have incomplete dominance over the other.)

The Bridge
answers

12. $x^{30}x^4 = x^{34}$

This is true since $x^{30}x^4 =$ xxxxxxxxxxxxxxxxxxxxxxxxxxxxxx xxxx $= x^{34}$

13. 16 cubic inches   If the diameter = 4, then r = 2.

$V = \frac{1}{3}\pi r^2 h = (\frac{1}{3})(3)(2^2)(4) = 16$

14. $\frac{3}{4} \times \frac{3}{4} \times \frac{3}{4} = \frac{27}{64}$

15. Yes. Integument = outer covering. Your skin, hair, and nails are part of your integumentary system.

16. 40% of 0.07 grams is 0.028 grams

17. 6   AA, AB, AO, BB, BO, and OO.

18. Fat.

19. B   Your basal metabolic rate occurs when your life processes are near their minimums. (Dead people don't have life processes, so D is not the correct answer.)

20. 17.3 – 1.7 = 15.6 pounds

21.
```
         90
        180      one double
        360      two doubles
        720      three doubles
      1440       four doubles
      2880       five doubles
      5760       six doubles
     11,520      seven doubles
     23,040      eight doubles
     46,080      nine doubles
     92,160      ten doubles
    184,320      eleven doubles
    368,640      twelve doubles
    737,280      thirteen doubles
  1,474,560      fourteen doubles
```

from p. 262 — *fifth try*

1. Since $\pi \approx 3.14$, it is true that $2\pi > 6$.

2. $2C_2H_4O + H_2 \rightarrow C_4H_6 + 2H_2O$

3. $\frac{49}{4}$

The Bridge
answers

4. Original equation $\quad \dfrac{27}{2} = \dfrac{10}{w}$

   Cross multiply $\quad 27w = 20$

   Divide both sides by 27 $\quad w = \dfrac{20}{27}$

5. $\dfrac{1}{8} = 12\tfrac{1}{2}\%$

6. It has one term. (see the last example on the fourth page of Chapter 25)

7. The three things that the dermis makes which poke through the epidermis are (1) hair, (2) sweat glands, and (3) oil glands.

8. We have 46 chromosomes. (23 pairs)

9. $y^6 y^6 = y^{12}$
   This is true because $y^6 y^6 = yyyyyy\ yyyyyy = y^{12}$

10. Original equation $\quad 66 + 8x + 87 = 3x + x + 333$

    Combine like terms $\quad 8x + 153 = 4x + 333$

    Subtract 4x from both sides $\quad 4x + 153 = 333$

    Subtract 153 from both sides $\quad 4x = 180$

    Divide both sides by 4 $\quad x = 45$

11. $0.5 \times 0.5 = 0.25$ (Or in fractions, $1/2 \times 1/2 = 1/4$)

12. Our brain

13. Photosynthesis

14. $6,000,000   (or $6,000,000.00)

15. Zero. Any number times zero is always equal to zero.

16. 81 cubic inches $\quad 9 \times 6 \times 1.5 = 81$

    (or, if you worked in fractions, $\dfrac{9}{1} \times \dfrac{6}{1} \times \dfrac{3}{2} = \dfrac{9}{1} \times \dfrac{\cancel{6}^{3}}{1} \times \dfrac{3}{\cancel{2}_{1}} = 81$)

17. Using conversion factors: $\dfrac{14\ \text{oz.}}{1} \times \dfrac{1\ \text{qt.}}{32\ \text{oz.}} \times \dfrac{1200\ \text{calories}}{1\ \text{qt}} = 525$ calories

18. $9.60   There are two ways to do this problem. One way is to first compute 20% of $8. ($0.20 \times 8 = 1.60$) and then add the $1.60 to the $8.

    The shorter way: A 20% gain means the original income (100%) plus 20%.

    $100\% + 20\% = 120\%$ 120% of $8 = 1.2×8 = $9.60

19. $20\% = \dfrac{20}{100} = \dfrac{1}{5}$

20. B) his retina which are the photoreceptors on the back wall of the eyes.

21. 15 is a cardinal number.

Index

air—its composition...... 87-89
alimentary canal............ 96
alleles................... 242
alliteration................ 38
antiperspirants........... 219
aorta.................... 151
approximately equal to ≈
 .... 25, 202
area of a rectangle.......... 37
argument by contradiction... 186
audioreceptors............ 100
autotrophs................. 97
Avogadro, Amedeo........ 166
Avogadro's number........ 167
basal metabolic rate........ 147
billion................ 94, 163
blood pressure............ 128
blood type................ 245
bones—how many?........ 201
brain—oxygen use......... 99
breast cancer............. 204
breathing in.............. 146
budget—making a......... 43
butane................... 175
calcium needs............. 203
calculators................. 13
Cantor counts to infinity.......
 .... 232, 233
capillaries........... 122, 123
cardinal numbers........... 48
cartilage................. 201
cell...................... 95
chemoreceptors........... 100
chlorophyll........... 160, 161

chromosomes............. 231
 naming system........ 234
circle
 circumference........... 41
 diameter............... 41
circular definition........... 30
circumference............. 41
class..................... 70
codominant alleles........ 245
color blindness........... 241
coma.................... 126
combining like terms....... 131
Concord grapes........... 246
conversion factor............
 .... 75, 76, 174, 240
Corleone, Don Vito....... 211
corn in Scotland........... 71
cowboys and hair conditioner...
 221
cranium................. 200
cri du chat............... 226
cross multiplying...... 186-188
crustaceans............... 161
cubed................... 133
cystic fibrosis........ 238, 243
d = rt................. 58, 59
dandruff................. 211
deodorants............... 219
dermis............... 216, 217
Divine Comedy............ 30
diameter.................. 41
dimensional analysis........ 59
dingbat.............. 241, 242
diploid.................. 231
discounts................. 64
disjoint sets............... 27
dividing fractions........... 32
divisible by 2............. 183

divisible by 3. 183
divisible by 5. 183
division by zero. 185, 186
DNA. 238
dominant gene. 110
Double R rule. 109
doubling and re-doubling. . . . . . .
 213-215
Down syndrome. 233
Drosophila. 222, 223, 229
Eddington, Sir Arthur. 77
effect of reading good books. . . .
 107
electroreceptors. 100
epiphany. 79
esophagus. 95
exponents. 133
facts of life concerning foods. . . .
 126
fainting. 73, 74
family. 70
femur. 200
fish—how they breathe. 147
flu shot. 243
fogging a mirror. 151
fontanels. 7
fractions
 adding. 24
 common denominator. 54
 dividing. 32
 multiplying. 32
Galen. 120-122
gene. 108
 dominant. 110
 recessive. 113
General Rule. 68, 74,
 75, 77, 152, 157,182
general rules for fractions—
 four of them. 185
genetics. 109
genotypes. 109, 227

genus. 70
germination of seeds. 37
giant hint #1 for stoichiometry. . .
 171
giant hint #2 for stoichiometry. . .
 172
glucose. 161
gram. 100
gustation. 100
hair. 218
Harvey, William. 119-122
the heart of biology. 79, 80
hemocyanin. 161
hemoglobin. 159-161
heterotrophs. 97
hexaploid. 231
Hsu, T. C. 232
Human Genome Project. . . . .
 . . . . 239, 243
hyperbole. 220
hypertension. 127
hypothalamus. 100
i before e, except after c rule. . . .
 127
ice cream for dogs. 157
improper fraction. 32
incisors. 95
incomplete dominance. 245
indirect proofs. 186
infinitely rich. 164
informal, general, and formal
 English. 207
integument. 207, 208
irony. 102, 105
junk DNA. 243
Kansas Fair Contest Law. 65
kingdoms—five of them. . 69, 70
Levan, Albert. 232
lien on property. 105, 150
life cycles. 44, 45
Life of Fred: Eating Pizza. . . . 11

life—a definition. 27-29
Linnaeus, Carolus.. . . . . 67, 69
liter. 96
litotes. 126, 127
locus. 225, 238
Logic. 13
Mae West. 12
Malpighi, Marcello. . . . 122, 123
mammal. 218
mechanoreceptors.. 101
Mendel, Gregor. 109, 244
metaphor. 159
meter. 96
mitosis. 223
mitotic division. 210
mixed number. 32
molars.. 95
mole—five definitions. 167
molecular weight. 166-168
mucus. 146
multiplying by 100. 36
multiplying by zero. 36
multiplying fractions. 32
mutant. 222
mutation. 222
    point mutations. 226, 242
    random. 225
    rare. 225
    regular. 224
Name Your Favorite Mutation
    game. 246
nefarious plans. 62
negative number. 106
neurons.. 208
nine hairy complications
    of biology. 243-246
nucleotides. 239
octaploid.. 231
oil glands. 219
olfaction. 100
order. 70

ordinal numbers. 40
osteoporosis. 204
parallel postulate of geometry
. . . . 89
patella.. 200
pharynx.. 95
phenotype. 110, 227
phenotype affecting genotype?
. . . . 245, 246
Photoreceptors. 101
photosynthesis. 91-93, 97
phylum. 70
pi. 41
    approximations. 41
pinna.. 100
plaque.. 55
pleiotropic. 244
point mutations. 242
polygenic. 109
power. 133
Prof. Eldwood's *A Guide to*
    Things Biological. . . . 230
Prof. Eldwood's *Digestion—Life*
    Along the Alimentary
    Canal. 96
Prof. Eldwood's *A Chicken Is*
    the Way an Egg Makes
    Another Egg. 44
Prof. Eldwood's First-Aid Quiz
. . . . 74
Prof. Eldwood's *Guide to*
    Fainting. 73
Prof. Eldwood's *Guide to Pizza*
. . . . 30
Prof. Eldwood's *Heart of*
    Biology. 91
Prof. Eldwood's *Magazine for*
    Future Brides. 51
proprioceptors. 101
quad—three meanings. 130
quadrillion. 94, 163-165

reading speed. 14
recessive gene. 113
reduction. 202
retina. 101
retroviruses. 243
ribs in men and women. 201
right angle. 135
right triangle. 135
rounds off to ≐ 124
saliva. 95
salivary amylase. 95
sebaceous glands. 219
sets
    disjoint. 27
    subset. 90
    union. 26
sibilant. 14
simile. 159
site. 242
skeleton equation. 161
smoker's broken bones. 205
smoking. 178
solipsism. 198, 201
species. 70
spelling rules for kingdom . . .
    species. 70
sternum. 200
stoichiometry. 170
stork. 236
story problems. 141-143
subset. 90
sucrose. 173
sweat glands. 218
sweet and greasy. 177
systolic. 128
Tenth Commandment. 108
terms—counting them. 132
tetraploid. 231
thermoreceptors. 100
Tijo, Jo Hin. 232
trachea. 152

trading genes. 244
transposon insertions. 243
trillion. 94, 163
triploid. 231
unit analysis. 59
van Helmont, Jan Baptist. . . . 81
victory gardens. 33-35
volume of a cone. 40
volume of a cube. 38
volume of a cylinder. . . 118, 134
Watson and Crick. 238
Western cannon. 30
Why We Brush—the movie. . . 55
windpipe. 152
X and Y chromosomes. . . . .
. . . . 236, 237
Zea mays. 67, 68, 70
zillion. 135

If you would like to learn more about books written about Fred . . .

FredGauss.com